140
Advances in Polymer Science

AF780152

Springer-Verlag Berlin Heidelberg GmbH

Progress in Polyimide Chemistry I

Volume Editor: H. R. Kricheldorf

With contributions by
J. de Abajo, J. G. de la Campa, D. L. Dunson,
K.O. Gaw, J. L. Hedrick, Y. Imai, M. Kakimoto,
J. E. McGrath, S. J. Mecham, P. Mison, B. Sillion

Springer

This series presents critical reviews of the present and future trends in polymer and biopolymer science including chemistry, physical chemistry, physics and materials science. It is addressed to all scientists at universities and in industry who wish to keep abreast of advances in the topics covered.

As a rule, contributions are specially commissioned. The editors and publishers will, however, always be pleased to receive suggestions and supplementary information. Papers are accepted for „Advances in Polymer Science“ in English.

In references Advances in Polymer Science is abbreviated Adv. Polym. Sci. and is cited as a journal.

Springer WWW home page: http://www.springer.de

ISSN 0065-3195
ISBN 978-3-662-15954-5 ISBN 978-3-540-49815-5 (eBook)
DOI 10.1007/978-3-540-49815-5

Library of Congress Catalog Card Number 61642

Originally published by Springer-Verlag Berlin Heidelberg New York in 1999
Softcover reprint of the hardcover 1st edition 1999

Typesetting: Data conversion by MEDIO, Berlin
Cover: E. Kirchner, Heidelberg
SPIN: 10648290 02/3020 - 5 4 3 2 1 0 - Printed on acid-free paper

Volume Editor

Professor
Dr. H. R. Kricheldorf
Inst. für Technische und
Makromolekulare Chemie
Universität Hamburg
Bundesstraße 45
D-20146 Hamburg, Germany
kricheld@chemie.uni-hamburg.de

Editorial Board

Preface

Over the past four decades polymers containing imide groups (usually as building blocks of the polymer backbone) have attracted increasing interest of scientists engaged in fundamental research as well as that of companies looking into their application and commercialization. This situation will apparently continue in the future and justifies that from time to time reviews be published which sum up the current state of knowledge in this field. Imide groups may impart a variety of useful properties to polymers, e. g., thermal stability chain stiffness, crystallinity, mesogenic properties, photoreactivity etc. These lead to a broad variety of potential applications. This broad and somewhat heterogeneous field is difficult to cover in one single review or monograph. A rather comprehensive monograph was edited four years ago by K. Mittal, mainly concentrating on procedures and properties of technical interest. Most reviews presented in this volume of Advances in Polymer Science focus on fundamental research and touch topics not intensively discussed in the monograph by K. Mittal. Therefore, the editor of this work hopes that the reader will appreciate finding complementary information.

Finally I wish to thank all the contributors who made this work possible and I would like to thank Dr. Gert Schwarz for the revision of the manuscripts of the contributions 3 and 4.

Hamburg, September 1998 Hans R. Kricheldorf

Contents

Contents of Volume 141

Progress in Polyimide Chemistry II

Volume Editor: H. R. Kricheldorf

Rapid Synthesis of Polyimides from Nylon-Salt-Type Monomers

Yoshio Imai[1)]

Department of Organic and Polymeric Materials, Tokyo Institute of Technology, Meguroku, Tokyo 152, Japan

This paper reviews our recent findings on the rapid synthesis of polyimides from nylon-salt-type monomers composed of diamines and tetracarboxylic acids. Aromatic polyimides are generally prepared by a two-step procedure from aromatic diamines and aromatic tetracarboxylic dianhydrides. However, the starting point of polyimide synthesis dates way back to the mid-1950s, when aliphatic-aromatic polyimides were synthesized by the melt polycondensation of nylon-salt-type monomers. Recently we have recovered the lost "salt monomer method" that is very useful for the rapid synthesis of aliphatic-aromatic polyimides. In addition, the salt monomer method coupled with high-pressure polycondensation was found to be versatile for the facile preparation of aliphatic-aromatic polyimides having well-defined structures. The properties of a series of newly synthesized aliphatic-aromatic polyimides and the application of the salt monomer method are also included in this review.

Keywords. Rapid polyimide synthesis, Nylon-salt-type monomers, Solid-state thermal polycondensation, High pressure polycondensation, Microwave-induced polycondensation, Aromatic polyimides, Aliphatic-aromatic polyimides, Polyimide-carbon black composites, Polyimide silica hybrid materials

1 Present address: 1-9-2-303, Nakamagome, Ohta-ku, Tokyo 143, Japan

List of Symbols and Abbreviations

XPMA	salt monomers of aliphatic diamines (X stands for the number of methylene unit) and pyromellitic acid (PMA)
XPME	salt monomers of aliphatic diamines (X) and pyromellitic acid half diethyl ester (PME)
XBPA	salt monomers of aliphatic diamines (X) and 3,3',4,4'-biphenyltetracarboxylic acid (BPA)
XTPE	salt monomers of aliphatic diamines (X) and 3,3",4,4"-*p*-terphenyltetracarboxylic acid half diethyl ester (TPE)
XOPA	salt monomers of aliphatic diamines (X) and 4,4'-oxydiphtalic acid (OPA)
XBTA	salt monomers of aliphatic diamines (X) and 3,3',4,4'-benzophenonetetracarboxylic acid (BTA)
ODPMA	salt monomer of bis(4-aminophenyl) ether (ODA) and pyromellitic acid (PMA)
ODPME	salt monomer of bis(4-aminophenyl) ether (ODA) and pyromellitic acid half diethyl ester (PME)
P-XPM	polyimides from salt monomers XPMA and XPME
P-XBP	polyimides from salt monomers XBPA
P-XTP	polyimides from salt monomers XTPE
P-XOP	polyimides from salt monomers XOPA
P-XBT	polyimides from salt monomers XBTA
P-ODPM	polyimides from salt monomers ODPMA and ODPME
DTA	differential thermal analysis
DSC	differential scanning calorimetry
TG	thermogravimetry
Tg	glass transition temperature
Tm	polymer melting temperature

1 Introduction

Aromatic polyimides are most useful super engineering plastics which exhibit excellent thermal, electrical, and mechanical properties, and have been used widely in aerospace, electronics, and other industries over the past three decades [1–4]. Aromatic polyimides are generally prepared through a two-step procedure by the ring-opening polyaddition of aromatic diamines to aromatic tetracarboxylic dianhydrides in NMP (or DMAc) solution giving soluble polyamic acids, followed by thermal cyclodehydration (Eq. 1) [1–5].

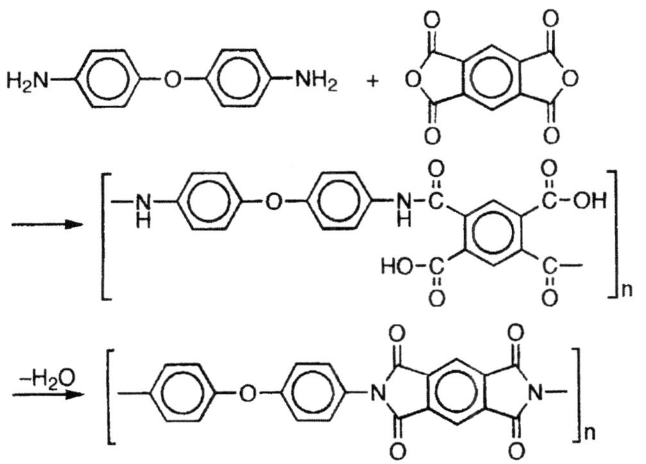

(1)

However, the starting point of polyimide synthesis dates away back to the mid 1950s, when Edwards and Robinson had synthesized aliphatic-aromatic polyimides (hereafter referred to as aliphatic polyimides) by the melt polycondensation of nylon-salt-type monomers composed of aliphatic diamines and aromatic tetracarboxylic acids or their diacid-diesters (Eq. 2) [6, 7].

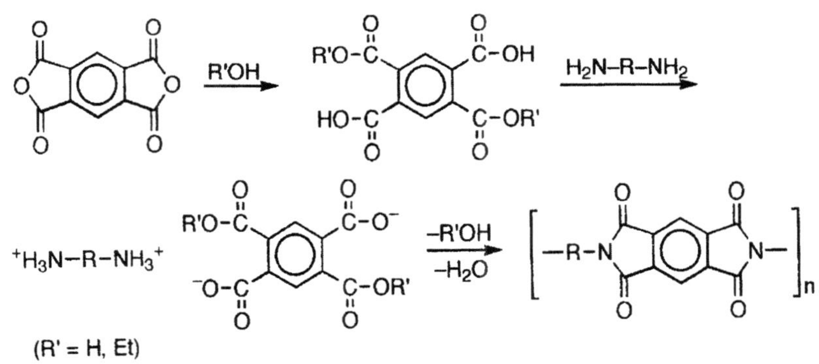

(2)

Later in 1967, an attempt was made to synthesize aromatic polyimide by the thermal polycondensation of the salt monomer derived from bis(4-aminophenyl) ether and pyromellitic acid half diester (Eq. 3, R'=isopropyl) [8].

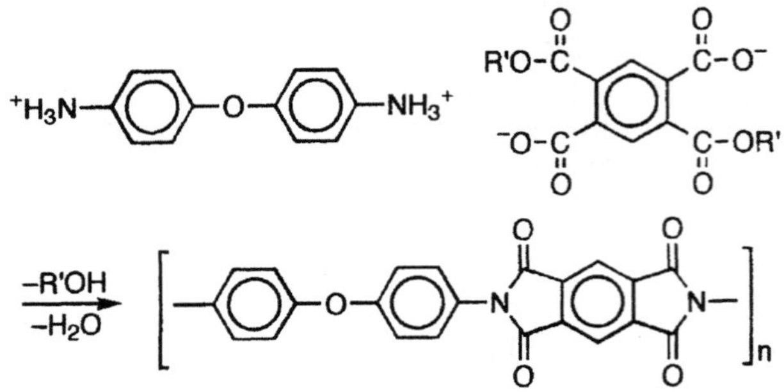

(3)

(R' = H, Et, Pr)

Recently Russian workers reported the kinetic and thermodynamic characteristics of thermal polycondensation of the salt monomers derived from some diamines and benzophenonetetracarboxylic acid half dimethyl ester leading directly to polyimides from the viewpoint of the preparation of polyimide-based advanced composite materials (Eq. 4, R'=methyl) [9–12].

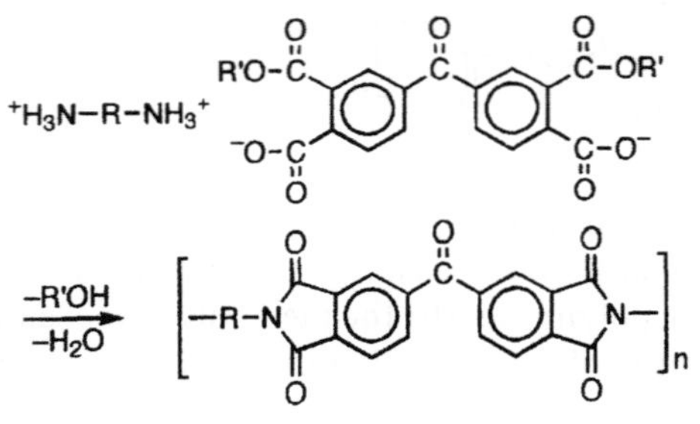

(4)

(R' = Me)

Nevertheless, further detailed information was unavailable on the polyimide synthesis from nylon-salt-type monomers that is referred to as "salt monomer method", and this method was not really recognized as a simple synthetic method of both aromatic and aliphatic polyimides. In addition, many polyimide investigations have mainly been concentrated on aromatic polyimides, and little information is available about aliphatic polyimides [13–18] that are also potential candidates for engineering plastics.

Quite recently we have found that the salt monomers were extremely reactive, producing directly polyimides in a very short reaction time (Eq. 5) [19]. In this connection, we have also found that the salt monomer method coupled with high pressure polycondensation was highly effective for the synthesis of aliphatic polyimides with well-defined structures.

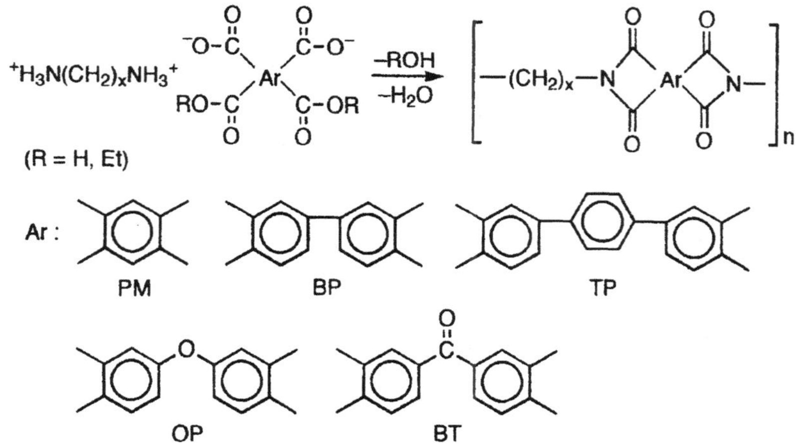

(5)

Thus, we have recovered the lost "salt monomer method" for a facile and versatile synthetic method for polyimides. This chapter reviews our recent findings on the rapid synthesis of a series of polyimides by the salt monomer method. This also includes the properties of newly synthesized aliphatic polyimides, and the application of the salt monomer method as well.

2 Solid-State Thermal Polycondensation of Salt Monomers

2.1 Preparation of Salt Monomers

First, the nylon-salt-type monomers such as XPME [20], XTPE [21, 22], and ODPME [23] with exactly 1:1 composition were prepared from a series of aliphatic diamines (X standing for the number of methylene unit) or aromatic bis(4-aminophenyl) ether (ODA) and the corresponding aromatic tetracarboxylic acid half diesters (diacid-diesters) such as pyromellitic acid diethyl ester (PME) and 3,3",4,4"-*p*-terphenyltetracarboxylic acid diethyl ester (TPE).

These salt monomers were prepared readily as white crystalline solids by dissolving an equimolar amount of each individual diamine and tetracarboxylic acid half diester in hot ethanol (or methanol), and subsequently cooling the resultant solution. The ease of preparation of the salt monomers is based on ready accessibility of the precise 1:1 stoichiometric balance due to the diamine-dicarboxylic acid functionalities, just like the well-known formation of nylon salts from aliphatic diamines and dicarboxylic acids.

Interestingly, aromatic tetracarboxylic acids themselves such as pyromellitic acid (PMA), 3,3',4,4'-biphenyltetracarboxylic acid (BPA), 4,4'-oxydiphthalic acid

(OPA), and 3,3',4,4'-benzophenonetetracarboxylic acid (BTA), coupled with both aliphatic and aromatic diamines, also led to the ready formation of the salt monomers having exactly 1:1 composition such as XPMA [24], XBPA [24], XOPA [25], XBTA [26], and ODPMA [27], despite possessing the diamine-tetracarboxylic acid functionalities.

2.2 Solid-State Thermal Polycondensation of Salt Monomers

The polycondensation of the salt monomers of both aliphatic-aromatic and wholly aromatic types proceeds according to Eqs. (5) and (3), giving aliphatic and aromatic polyimides, respectively.

Prior to the polycondensation, the thermal behavior of the salt monomers was investigated in some detail. Figure 1 shows the DTA and TG curves of salt monomer 12PMA consisting of dodecamethylenediamine and pyromellitic acid, which are typical for the aliphatic-aromatic salt monomers (see Eq. 5, X=12, Ar=PM, and R=H) [24].

The DTA thermogram exhibited a sharp endotherm at 199 °C, and the TG curve recorded 13.5% weight loss starting at around 170 °C and ending at 220 °C. The weight loss value observed was in close agreement with the calculated value of 13.6%. The calculation is based on the fact that during the imide-forming reaction from the salt, four molar equivalents of water should be lost. These results clearly indicated that the DTA endotherm was not the real melting point of the salt monomer but the imide-forming reaction temperature. Hence, it could be

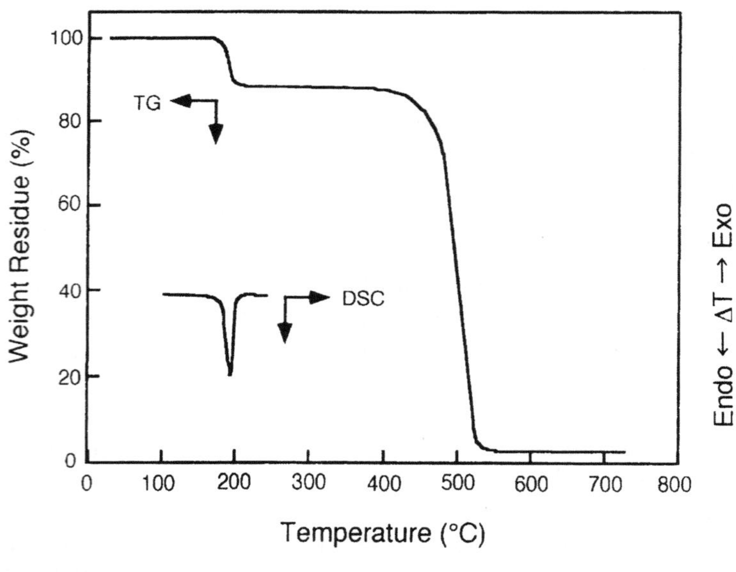

Fig. 1. DTA and TG curves of salt monomer 12PMA at a heating rate of 10 °C/min in air

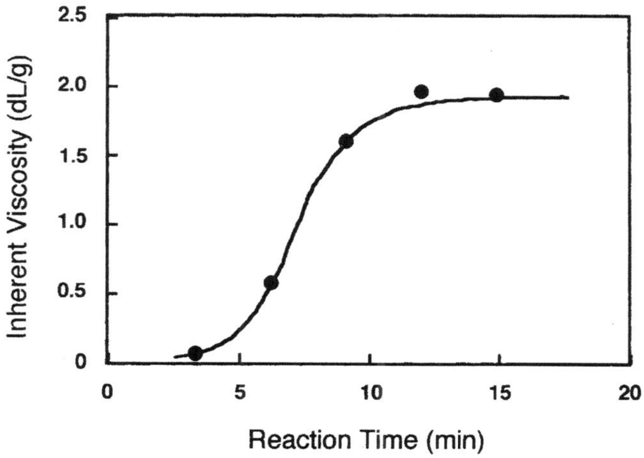

Fig. 2. Time dependence of inherent viscosity of polyimide P-12PM formed by the solid-state polycondensation of salt monomer 12PMA at 250 °C

concluded that the aliphatic-aromatic salt monomer was highly reactive and converted rapidly to the aliphatic polyimide (Tm=320 °C) in a solid state during such a short time for the thermal analysis, apparently through the one-step direct imide-forming reaction rather than the conventional two-step reaction via polyamic acid that is a precursor of polyimide.

Figure 2 exhibits the time dependence of the inherent viscosity of polyimide P-12PM formed by the solid-state thermal polycondensation of salt monomer 12PMA [28]. Surprisingly, the polycondensation proceeded rapidly at 250 °C, and was almost complete within only 10 min, affording the aliphatic polyimide having high inherent viscosity around 2.0 dL/g, despite the fact that the polymerization temperature was far lower than the melting temperature of the polyimide of 320 °C. Such an ease of the polycondensation is attributable to high reactivity of the salt monomer.

Another successful example is the solid-state thermal synthesis of the wholly aromatic polyimide P-ODPM (Tg=410 °C) by the polycondensation of the aromatic salt monomer ODPMA derived from bis(4-aminophenyl) ether and pyromellitic acid (see Eq. 3, R'=H). Figure 3 shows the DSC and TG curves of salt monomer ODPMA [27].

The DSC curve exhibited a sharp endotherm at 220 °C, and the TG curve recorded 15.2% weight loss starting at that temperature and ending at around 250 °C. The weight loss value observed agreed quite well with the calculated value of 15.9% based on the loss of four molar equivalents of water through the imide-ring formation from the salt. Therefore, the endotherm in the DTA curve is not the real melting point of the salt but is accompanied by the rapid direct formation of the polyimide with the elimination of water. These results indicated that, like the aliphatic-aromatic salt monomer 12PMA, the aromatic salt mono-

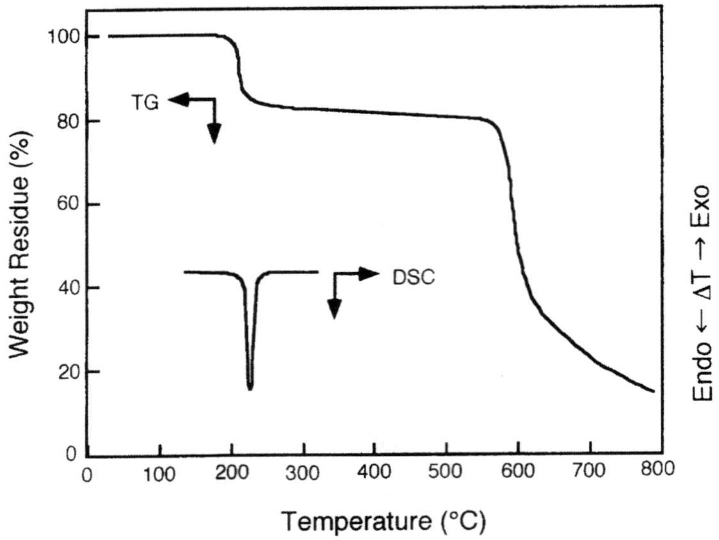

Fig. 3. DSC and TG curves of salt monomer ODPMA at a heating rate of 10 °C/min in air

mer ODPMA was also converted to the aromatic polyimide so rapidly at such the low temperature range around 220 °C, apparently by the one-step direct imide-forming reaction rather than by the usual two-step reaction via polyamic acid (see Eq. 1). Salt monomer ODPMA derived from bis(4-aminophenyl) ether and pyromellitic acid itself was somewhat less reactive than salt ODPME composed of the aromatic diamine and pyromellitic acid half diethyl ester from a comparison of the individual DSC curves [23].

The reaction temperature and time are important factors for the synthesis of the aromatic polyimide. Figure 4 shows the time dependence of inherent viscosity of aromatic polyimide P-ODPM synthesized by the polycondensation of aromatic salt monomer ODPMA at 240 °C [27].

Actually, the polycondensation proceeded rapidly at 240 °C and was almost complete within only 1 h. Even under these mild reaction conditions, the aromatic polyimide formed had an inherent viscosity around 0.7 dL/g, which was indicative of moderately high molecular weight. As mentioned above, the solid-state thermal polycondensation of the aliphatic diamine-based salt monomers like 12PMA was complete within 10 min [28], and hence the progress of the solid-state thermal polycondensation of the aromatic salt monomer ODPMA was slightly slower than that of the aliphatic diamine-based salt monomers. Nevertheless, such the rapid and direct formation of the aromatic polyimide having too high glass transition temperature of 410 °C at such a low reaction temperature of 240 °C in a solid state is particularly surprising, considering the aromatic polyimide synthesis through a general two-step procedure that requires higher temperature and longer reaction time.

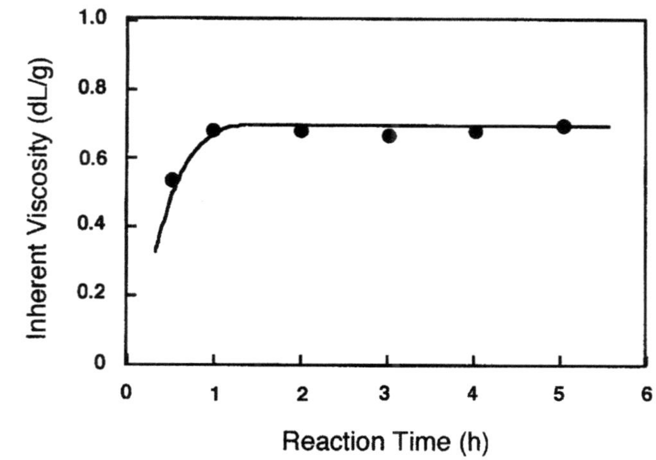

Fig. 4. Time dependence of inherent viscosity of polyimide P-ODPM formed by the solid-state polycondensation of salt monomer ODPMA at 240 °C

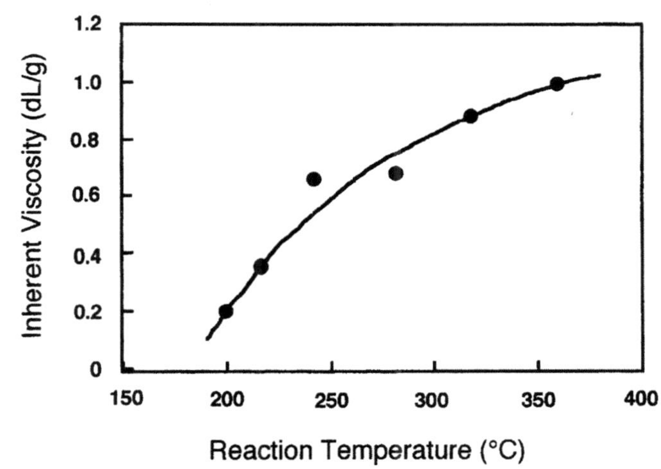

Fig. 5. Temperature dependence of inherent viscosity of polyimide P-ODPM formed by the polycondensation of salt monomer ODPMA for 1 h

Figure 5 represents the temperature dependence of inherent viscosity of the polyimide prepared by the polycondensation of salt monomer ODPMA for 1 h [27]. From the DSC and TG studies, the aromatic salt monomer was found to undergo polycondensation at the endothermic peak temperature of 220 °C. The polymerization at a temperature lower than 220 °C resulted in the polymer with low viscosity value. The inherent viscosity of the polymer increased with in-

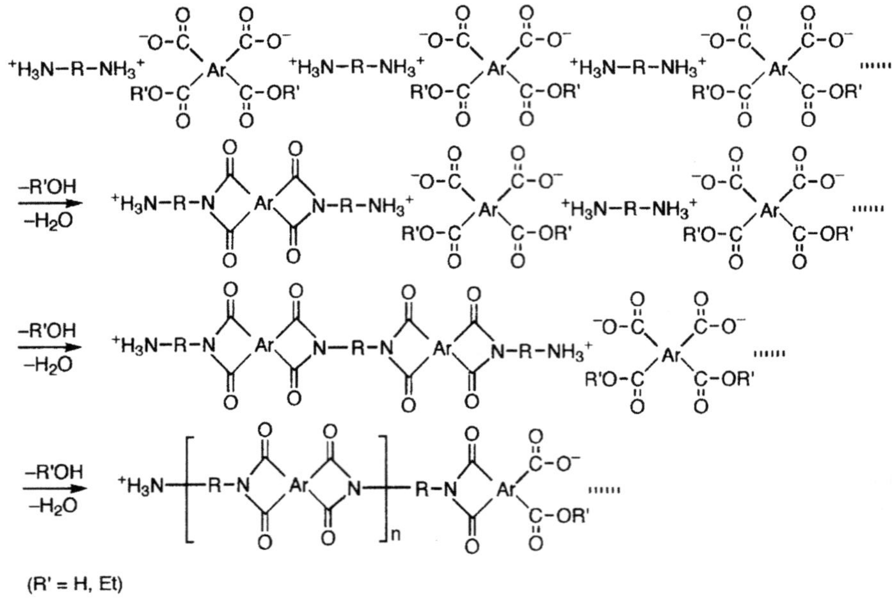

Scheme 1

creasing reaction temperature, and the high viscosity value over 1.0 dL/g was reached by the polycondensation at 360 °C for 1 h.

Such easy solid-state thermal polycondensation of the aromatic salt monomer under these mild conditions giving directly the wholly aromatic polyimide with high molecular weight is again remarkable. All the above results are presumably attributable to the fact that the salt monomers, both aliphatic-aromatic and all aromatic, are just simple nylon-salt-type monomers having exact diamine-tetracarboxylic acid (1:1) stoichiometric balance, which is the first requisite for the formation of high-molecular-weight condensation polymers. In addition, these results are ascribable to the fact that the reactive ion pairs are adjacently arranged in the solid salt crystals and the growing polymer chain ends, which facilitate the polyimide formation, as depicted in Scheme 1. Hence the salt monomers are quite different from usual AA-BB-type two component systems that need more energy for the melt polycondensation.

In a brief summary, we have developed a facile and versatile "salt monomer method" for the rapid synthesis of both aliphatic and aromatic polyimides. The salt monomer method has the following advantages over the conventional two-step method. First, the aliphatic-aromatic salt monomers as well as all aromatic, composed of diamines (both aliphatic and aromatic) and aromatic tetracarboxylic acids (or their half diesters) are highly reactive and rapidly produce polyimides with high molecular weights in one step by the solid-state thermal polycondensation. Second, this method is a facile and versatile synthetic method for polyimides from salt monomers, and is immediately applicable to simple laboratory synthesis of various polyimides. Third, the method is also promising for reactive processing such as reactive extrusion giving directly polyimide pellets, and reactive pulltrusion or reactive injection into polyimide moldings.

3 High-Pressure Polycondensation of Salt Monomers

3.1 High-Pressure Synthesis of Aliphatic Polyimides

A variety of methods are known for the synthesis of polyimides and other condensation polymers, however, the application of high pressure has seldom appeared in the literature to date. Early in 1969 Morgan and Scott reported on the high-pressure polycondensation and simultaneous hot-pressing of intractable polybenzimidazopyrrolone, that is infusible and insoluble, directly from the combination of 3,3',4,4'-tetraaminobiphenyl and pyromellitic dianhydride (Eq. 6) [29, 30].

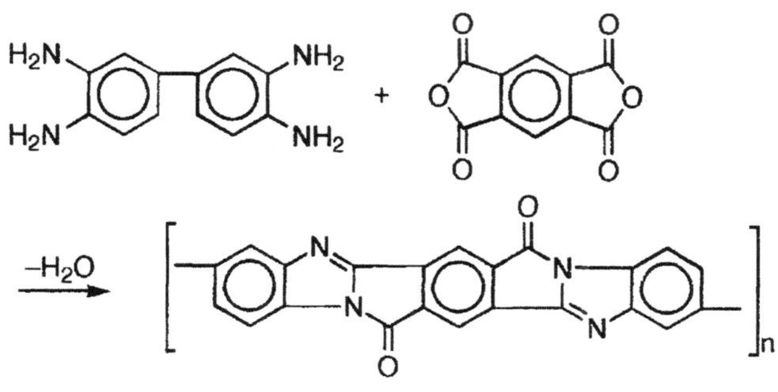

(6)

Later, Ikawa et al. studied the the high-pressure synthesis of aliphatic polyamides such as 11-nylon and 12-nylon starting from the corresponding ω-amino acids (Eq. 7) [31, 32].

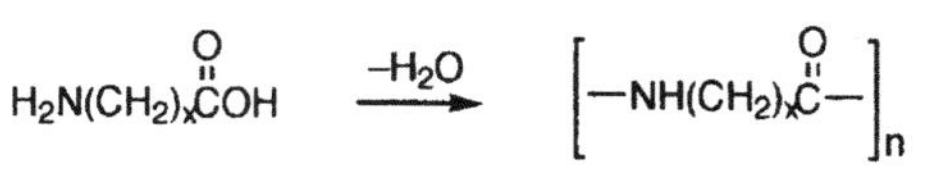

(7)

Quite recently, we have investigated systematically the high-pressure polycondensation leading to the formation of a variety of polyimides, and polybenzoxazoles as well [19, 33]. Here we applied the salt monomer method to the high-pressure synthesis of aliphatic polyimides.

In practice, the high-pressure polycondensation of the salt monomers producing polyimides was carried out by using a piston-cylinder type hot-pressing apparatus with the use of a Teflon capsule as a reaction vessel (see Eq. 5) [34].

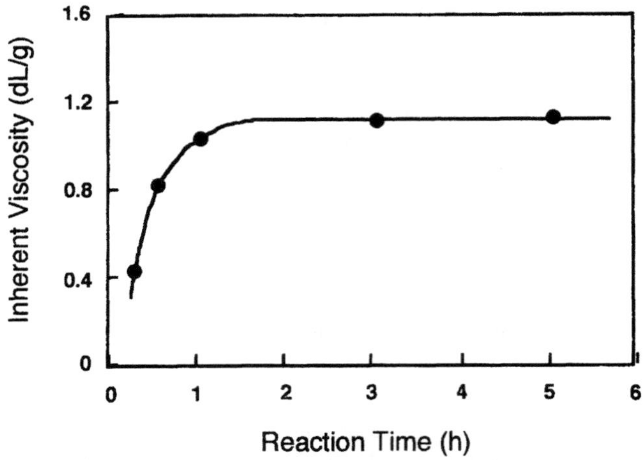

Fig. 6. Time dependence of inherent viscosity of polyimide P-12PM formed by the high-pressure polycondensation of salt monomer 12PMA under 220 MPa at 320 °C

Figure 6 shows the progress of the high-pressure polycondensation of salt monomer 12PMA under 220 MPa (about 2200 atmosphere) at 320°C, giving polyimide P-12PM (see Eq. 5, X=12, Ar=PM, and R=H) [24].

Since the melting temperature (Tm) of the resulting polyimide was 320°C, the Tm under the experimental high-pressure conditions should exceed well 320°C, and hence the polycondensation should proceed in a solid state. The polycondensation was found to proceed rapidly even under the high-pressure conditions and was almost complete within 1 h, giving the polyimide with high inherent viscosity above 1.0 dL/g, which was indicative of high molecular weight.

The result of the high-pressure polycondensation of the salt monomer that the salt directly afforded high-molecular-weight polyimide within such a short reaction time was particularly surprising in view of the facts that the by-product of water produced during the polycondensation co-existed in the closed reaction vessel, and the application of high pressure coupled with solid-state reaction system actually retarded the polycondensation due to restriction of mobility of the salt monomer and the growing polymer chain ends. Actually, some extent of moisture was detected on the surface of the polyimide pellet taken out of the closed reaction vessel just after the high-pressure polycondensation. Therefore, it is thought that the water produced was eliminated exclusively from the solid polymer mass throughout the reaction, namely, the high-pressure polycondensation proceeded rapidly under unequilibrium state. Such an ease of the high-pressure polycondensation is presumably attributable again to the fact that the reactive ion pairs are adjacently arranged in the salt monomer crystals and the growing polymer chain ends, similarly to the above-mentioned solid-state thermal polycondensation (see Scheme 1).

Table 1. High-pressure synthesis and thermal behavior of aliphatic polyimides

Salt monomer		Polymn conditions[b]	Polyimide			
Code	Tp[a] (°C)		Code	η_{inh}[c] (dL/g)	Tg[d] (°C)	Tm[d] (°C)
6PMA	258	A	P-6PM	0.62	–	–
7PMA	216	A	P-7PM	1.63	–	348
8PMA	274	A	P-8PM	0.39	–	374
9PMA	271	A	P-9PM	1.28	–	318
10PMA	245	A	P-10PM	1.08	–	347
11PMA	245	A	P-11PM	2.23	–	299
12PMA	248	A	P-12PM	1.17	–	321
6BPA	219	A	P-6BP	0.70	145	350
7BPA	224	A	P-7BP	1.53	131	260
8BPA	226	A	P-8BP	0.76	121	301
9BPA	209	A	P-9BP	1.75	108	231
10BPA	229	A	P-10BP	1.14	102	253
11BPA	202	A	P-11BP	2.72	91	222
12BPA	199	A	P-12BP	1.94	84	237
6TPE	185	B	P-6TP	0.60	141	352
7TPE	178	B	P-7TP	0.82	135	290
8TPE	172	B	P-8TP	1.20	120	304
9TPE	166	B	P-9TP	1.04	104	263
10TPE	165	B	P-10TP	0.58	103	291
11TPE	155,164	B	P-11TP	0.66	89	240
12TPE	166,210	B	P-12TP	0.66	86	271
6OP	204	C	P-6OP	0.47	118	230
7OP	221	C	P-7OP	1.56	110	–[f]
8OP	170,177	C	P-8OP	0.78	100	220
9OP	190	C	P-9OP	1.12	89	–[f]
10OP	202,217	C	P-10OP	2.89	83	175
11OP	193	C	P-11OP	1.77	74	–[f]
12OP	225	C	P-12OP	2.00	70	160
4BT	240	D	P-4BT	0.88	184	334
5BT	220	D	P-5BT	0.70	163	–[f]
6BT	187	D	P-6BT	1.00	145	267
7BT	175	D	P-7BT	–[e]	140	–[f]
8BT	206	D	P-8BT	1.38	122	244
9BT	205	D	P-9BT	–[e]	116	–[f]
10BT	223	D	P-10BT	0.95	105	200, 240
11BT	227	D	P-11BT	–[e]	100	170
12BT	211	D	P-12BT	0.45	95	202

[a] Endothermic peak temperature determined by DTA (or DSC) at a heating rate of 10 °C/min in nitrogen
[b] Polymerization conditions are as follows: A – under 220 MPa at 250 °C for 15 h; B – under 260 MPa at 250 °C for 15 h; C – under 220–300 MPa at 220–280 °C for 15 h depending on the salt monomers used; D – under 200–250 MPa at 200–320 °C for 5 h depending on the salt monomers used
[c] Inherent viscosity was measured at a concentration of 0.5 g/dL in concentrated sulfuric acid at 30 °C
[d] Tg and Tm values were determined by DSC at a heating rate of 10 °C/min in nitrogen
[e] The polyimide formed was insoluble in concentrated sulfuric acid
[f] No Tm was detected on the DSC curve

The results of the high-pressure polycondensations of the aliphatic-aromatic salt monomers XPMA, XBPA, XTPE, XOPA, and XBTA under 200–600 MPa at 220–320°C for 3–30 h leading to the formation of a series of aliphatic polyimides (see Eq. 5) are summarized in Table 1 [22, 24–26].

The high-pressure polycondensations readily afforded a series of linear aliphatic polyimides P-XPM, P-XBP, P-XTP, P-XOP, and P-XBT as pale yellow pellets having inherent viscosities as high as 2.8 dL/g. An additional feature of the high-pressure synthesis of polyimides is that the high-pressure polycondensation effectively induced crystallization of the growing polyimides, resulting in highly crystalline polyimides in most cases.

The reaction temperature and applied pressure strongly affected the molecular weight of the resulting polyimides. Under constant pressure, the inherent viscosity values of the polyimides increased with increasing reaction temperature. In contrast, at a constant temperature, the viscosity values decreased obviously as the applied pressure increased. Hence the application of high pressure to the polymerization system could not accelerate the polycondensation, but actually retarded. Thus, too much higher pressure is unfavorable for the high-pressure synthesis of polyimides with high molecular weights.

Contrary to the high-pressure polycondensation, when the polycondensation of the salt monomers was conducted in a molten state under atmospheric or reduced pressure for the preparation of the polyimides having Tm below 300°C, this often led to the formation of crosslinked aliphatic polyimides that were insoluble even in concentrated sulfuric acid. Therefore, the high-pressure polycondensation process provides a simple and effective method for the synthesis of the linear polyimides with well-defined structures that caused high crystallinity, compared with the other synthetic methods.

Then the reactivity of the tetracarboxylic acid-based salt monomers was compared with that of the salts consisting of tetracarboxylic acid half diesters. The P-XPM series polyimides were prepared by the high-pressure polycondensation of salt monomers XPME derived from the aliphatic diamines and pyromellitic acid half diethyl ester (see Eq. 5, Ar=PM and R=ethyl) [20], in addition to the polymers obtained from the pyromellitic acid-based salt monomer XPMA [24] already shown in Table 1. The polycondensation of salts XPME proceeded readily under high pressure of 250 MPa at 240 °C for 15 h, even with the elimination of ethanol and water as the by-products in the closed reaction vessel, and this afforded the polyimides with inherent viscosities up to 1.6 dL/g. Therefore, the reactivity of salt monomers XPME was found to be almost comparable to that of the parent salt XPMA, and the properties of the resultant P-XPM series polyimides from XPME were the same as those obtained from salts XPMA.

3.2 High-Pressure Synthesis of Aromatic Polyimides

The high-pressure synthesis of wholly aromatic polyimides directly from salt monomers would be promising in view of the reactive processing involving polymerization and simultaneous hot-pressing for intractable aromatic polyimides. The high-pressure polycondensation of salt monomer ODPME, composed of bis(4-aminophenyl) ether and pyromellitic acid half diethyl ester, was carried out under high pressure of 220 MPa at 280 °C for 5 h, in order to obtain aromatic polyimide P-ODPM (see Eq. 3, R'=ethyl) [23].

The polycondensation actually proceeded under these high-pressure conditions, although the inherent viscosity of the resultant polymer was 0.19 dL/g, indicative of low molecular weight. The polyimide oligomer thus formed had, of course, high crystallinity. When this oligomer was subjected to post-polymerization by heating at 400 °C for 1 h under atmospheric pressure, further polycondensation occurred rapidly, resulting in polyimide P-ODPM having high inherent viscosity of 1.1 dL/g. These results clearly indicated that to obtain the polyimide having high Tg of 410 °C, a high reaction temperature of 400 °C was essential to enhance the mobility of the living polymer chain ends consisting of salt ion-pair structure of the growing polyimide molecule, which is necessary for the progress of polycondensation in the solid state. The high-molecular-weight aromatic polyimide thus formed had also crystalline nature.

Furthermore, we have extended the high-pressure polycondensation to the synthesis of condensation polymers other than polyimides. A successful example was the high-pressure synthesis of polybenzoxazoles starting from the bis-*o*-aminophenol (or its hydrochloride) and aliphatic dinitriles (Eq. 8) [35]. The polycondensation readily proceeded under high pressure in a closed reaction vessel with the elimination of the by-product of ammonia (or ammonium chloride).

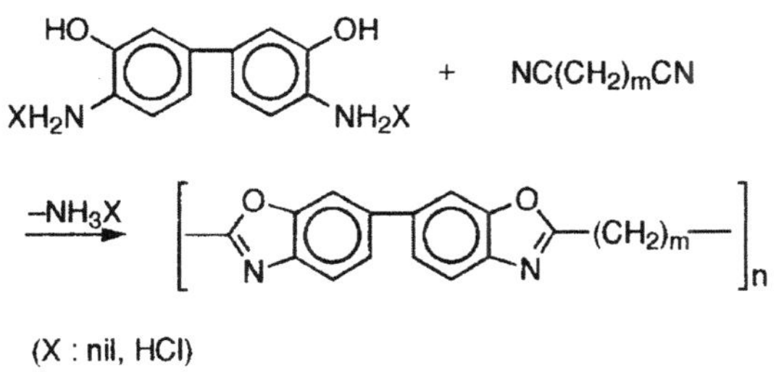

(8)

In addition to the high-pressure polycondensation reactions mentioned above, we have studied the high-pressure polyaddition for the synthesis of high-temperature aromatic polymers such as polyaminoimide from bismaleimide and aromatic diamine, polycyanurates from aromatic dicyanates, and polyisocyanurates from aromatic diisocyanates [34, 36–40].

3.3 Thermal Behavior of Aliphatic Polyimides

The thermal behavior of a series of newly synthesized aliphatic polyimides having well-defined structures, obtained by the high-pressure polycondensation of the salt monomers, was then studied. Table 1 summarizes the Tg and Tm values of the aliphatic polyimides.

Most of the aliphatic polyimides, except for those of the P-XBT and P-XOP series with odd number of methylene unit, were crystalline polymers having clear Tm. In particular, polyimide P-XPM were highly crystalline and had no Tg. The Tm values of a series of the aliphatic polyimides having 6–12 methylene units are strongly dependent on aromatic tetracarboxylic acid components, and decrease in the following order: P-XPM (Tm:>400–300 °C) [24]>P-XTP (350–240 °C) [22]>P-XBP (350–220 °C) [24]>P-XBT (265–200 °C) [26]>P-XOP (230–160 °C) [25]. A marked odd-even effect for the Tm values of the aliphatic polyimides was observed with respect to the number of methylene units, where the polyimides having even-numbered methylene units exhibited higher Tm values than their odd-numbered neighbors. The decreasing order for the Tg values of the aliphatic polyimides was almost the same as that for the Tm values: P-XTP (141–86 °C) [22]≈P-XBP (145–84 °C) [24]>P-XBT (145–95 °C) [26]>P-XOP (118–70 °C) [25], where the Tg values decreased monotonically as the number of methylene unit increased. These decreasing orders for the Tm and Tg values are consistent with the decreasing rigidity of the polyimide backbones.

From these results, it may be concluded that the aliphatic polyimides, especially the pyromellitic acid-based polyimides P-XPM with Tm values around 300 °C, are promising candidates as moldable high-temperature polyimide resins useful for industrial applications.

3.4 Thermotropic Liquid-Crystalline Aliphatic Polyimides

Recently much effort has been made to induce liquid crystallinity in polyimides. Kricheldorf et al., Chen et al., and Sato et al. have suggested that an aromatic imide ring may be an excellent mesogen through their studies on thermotropic liquid-crystalline polyimide-esters [41–54] and polyimide-carbonates [55, 56]. Some thermotropic properties of aromatic polyether-imides have also been reported by Asanuma et al. [57].

During the course of our studies on liquid crystalline polyimides of main-chain type, we have synthesized simple polyimides P-XPM and P-XBP, where the aromatic mesogen and the alkylene spacer are connected in an alternating fashion through the imide ring structure, by the high-pressure polycondensation of the salt monomers [24]. The pyromellitimide and biphenyltetracarboxydiimide groups in the polyimides should be stiff and long enough to form liquid crystals, since the imide rings can function as a part of the mesogen. However, so far we have not observed any mesophases in the main-chain type polyimides based on these mesogenic groups.

More recently, we have found that the terphenyltetracarboxylic acid-based polyimides P-XTP showed thermotropic liquid-crystalline phase [21, 22]. As far as we know, this is the first example of the simple polyimides showing thermotropic liquid crystallinity.

Among polyimides P-XTP, the polyimides with X=8–12 formed mesophases. The mesophases of P-8TP and P-11TP were enantiotropic, whereas those of the other polyimides were monotropic. The nematic phase was identified for all the polyimides from X-ray and microscopic observations. In particular, the DSC thermogram of polyimide P-11TP exhibited three distinct endothermic peaks upon heating, and again three exothermic peaks upon cooling, indicating that the polyimide formed the enantiotropic liquid crystal from 228 to 240 °C on heating and from 210 to 169 °C on cooling. Additionally, only polyimide P-11TP exhibited polymorphism, where a nematic phase was followed by a smectic phase upon cooling.

4
Microwave-Induced Polycondensation of Salt Monomers

The application of microwave energy using a domestic microwave oven is a promising methodology for polymer synthesis. Microwave energy has been utilized for the radical polymerization of vinyl monomers and for the curing reactions of epoxy resins and a polyamic acid to polyimide during the last decade. However, there has so far been no reports on the synthesis of condensation polymers with the use of microwave energy. We have developed a new microwave-induced polycondensation method for the facile and rapid synthesis of aliphatic polyimides and aliphatic polyamides as well [58, 59].

The microwave-induced polycondensation was carried out simply by heating monomers in the presence of a polar organic medium in a domestic microwave oven. The microwave energy effectively produces sufficient exothermic molecular friction of the reaction mixtures in solution rather than in solid state at an initial stage of the reaction, thus facilitating polycondensation. For the microwave-induced polycondensation, a polar solvent with high boiling point coupled with good solubility of the monomers must be used due to the high reaction temperature, usually around 300 °C, that is required for the reaction. The most suitable solvents for the polycondensation were found to be *N*-cyclohexyl-2-pyrrolidone (CHP) and 1,3-dimethyl-2-imidazolidone (DMI). Typically, the polycondensation started in solution and proceeded in a plasticized melt state after rapid evaporation of most of the solvent at high temperature.

Here we have combined the salt monomer method with microwave-induced polycondensation for the synthesis of aliphatic polyimides P-XPM from salt monomers XPMA and XPME (see Eq. 5, X=6–12, Ar=PM, and R=H and ethyl) [28]. When DMI or CHP was used as the solvent, the polycondensation of both salt monomers proceeded quite rapidly, and only 2 min of microwave irradiation readily afforded the aliphatic polyimides with inherent viscosities around 0.7 dL/g or above. Under these microwave irradiation conditions, salt monomers XPME were found to be more reactive than salts XPMA, judging from the attained inherent viscosity values.

The microwave-induced polycondensation was also applied to the preparation of aliphatic polyamides. When ω-amino acids and nylon salts were used as monomers, the polycondensation proceeded rapidly, giving aliphatic polyamides with inherent viscosities around 0.5 dL/g by only 4–5 min of microwave heating (Eqs. 7 and 9) [60, 61].

$$^{+}H_3N(CH_2)_xNH_3^{+}\ {}^{-}O\overset{O}{\overset{\|}{C}}(CH_2)_y\overset{O}{\overset{\|}{C}}O^{-} \xrightarrow{-H_2O} \left[-NH(CH_2)_xNH\overset{O}{\overset{\|}{C}}(CH_2)_y\overset{O}{\overset{\|}{C}}- \right]_n \quad (9)$$

In short, the microwave energy was successfully applied for the first time for the rapid synthesis of condensation polymers such as polyimides and polyamides from the salt monomers (and ω-amino acids) by using a simple domestic microwave oven. The rapid polymer formation is caused by the efficient internal heating of the reactants. This compares favorably with the long reaction time required for the conventional thermal polycondensation with external heating.

5 Application of Salt Monomer Method

5.1 Preparation of Electro-Conductive Polyimide-Carbon Black Composites

The salt monomer method was successfully applied to the preparation of the electrically-conducting polyimide-carbon black composites [62]. The composites are prepared as follows: An aqueous solution of salt monomer 9PMA was mixed with carbon black, giving a suspension. This was evaporated to dryness under reduced pressure to afford a homogeneously-mixed powder composed of the salt monomer and carbon black. The powder was subjected to solid-state thermal polycondensation in the form of a pellet at 240 °C for 1 h under atmospheric pressure. The semiconducting aliphatic polyimides (P-9PM, Tm=315 °C) having electric conductivity of about 10^{-6} S/cm was readily obtained by mixing only 1 wt% of carbon black based on the polyimide.

5.2 Preparation of High-Performance Polyimide-Silica Hybrid Materials

The high-pressure polyimide synthesis from salt monomers was extended to the development of a new approach for hybrid materials composed of aliphatic polyimides and silica, wherein this method was combined with the sol-gel process [63, 64]. The preparation process is outlined in Scheme 2, where aliphatic polyimide P-9PM was used as the polyimide component.

An aqueous solution of salt monomer 9PME was mixed with tetramethoxysilane, yielding a gel, which was then vacuum-dried forming a precursor powder composed of silica gel and the salt monomer. This was subjected to high-pressure polycondensation under 235 MPa at 230 °C for 5 h, producing polyimide-silica hybrid molding. By varying the ratio of tetramethoxysilane to the salt

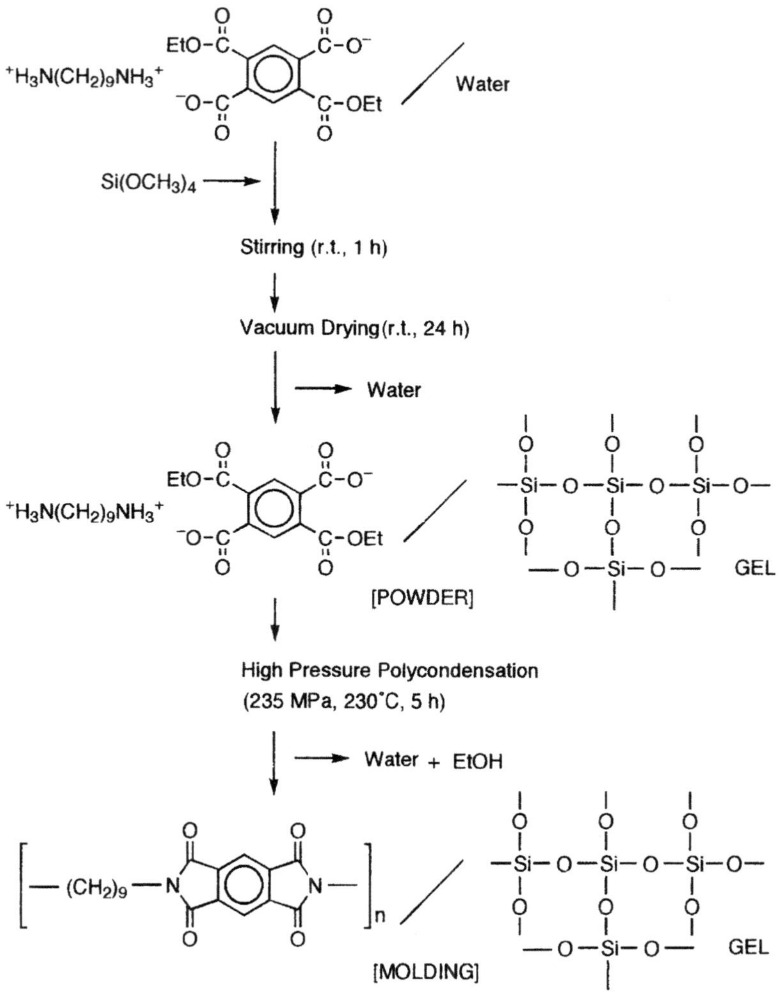

Scheme 2

monomer, the hybrid materials containing 10–100 wt% of silica could be prepared readily.

The hybrid materials having silica content below 50 wt% were composed of polyimide matrix with finely dispersed silica particles, and their hardness values were very close to that of the matrix polyimide. On the other hand, the hybrid materials having a silica content over 50 wt% were very hard and tough, and their hardness values increased with increasing silica content. In the latter hybrid materials, the silica formed a continuous phase with polyimide as the minor phase that probably acts as binder. This is a new type of polyimide-based composite, and may be referred to as "polyimide-reinforced silica glass", although we have already developed another method for the preparation of polyimide-silica hybrid materials as silica-reinforced polyimides through the sol-gel process [65–71].

6 Conclusions

We have developed a facile and versatile "salt monomer method" for the rapid synthesis of both aliphatic and aromatic polyimides. The solid-state thermal polycondensation of the nylon-salt-type monomers, composed of aliphatic diamines (or aromatic diamines) and aromatic tetracarboxylic acids or their half diacid-diesters, proceeded rapidly at around 250 °C within 10 min, affording aliphatic polyimides with high molecular weights. The rates of the polycondensations of the salt monomers under various conditions were in the following order: the microwave-induced polycondensation>the solid-state thermal polycondensation>the high-pressure thermal polycondensation. The most striking aspect of the polycondensation of the salt monomers is that the aliphatic-aromatic salt monomers as well as all aromatic are highly reactive and rapidly produced the polyimides with high molecular weights through the one-step direct imide-forming reaction rather than the conventional two-step process via polyamic acids. Furthermore, the high-pressure thermal polycondensation of the salt monomers is very useful for the synthesis of the polyimides having well-defined structures, compared with the other synthetic methods. Thus, the salt monomer method is applicable to simple laboratory synthesis of various polyimides, and also promising for reactive processing, such as reaction extrusion, reactive pulltrusion, reactive injection, and reactive compression, giving directly polyimide pellets and moldings. In addition, the aliphatic polyimides, especially the pyromellitic acid-based polyimides with Tm values around 300 °C, are potential candidates as moldable high-temperature polyimide resins useful for industrial applications.

7 References

1. Sroog CE (1969) Encycl Polym Sci Tech 11:247
2. Bessonov MI, Koton MM, Kudryavtsev VV, Laius LA (1987) Polyimides: thermally stable polymers. Consultants Bureau, New York
3. Wilson D, Stenzenberger HD, Hergenrother PM (1990) Polyimides. Blackie, New York
4. Sroog CE (1991) Prog Polym Sci 16:561
5. de Abajo J (1992) Polyimides. In: Kricheldorf HR (ed) Handbook of polymer synthesis, pt B. Dekker, New York, p941
6. Edwards WM, Robinson IM (1955) US Pat 2,710,853
7. Sorenson WR, Campbell TW (1968) Preparative methods of polymer chemistry, 2nd edn. Interscience, New York, p88
8. Bell VL (1967) J Polym Sci, Part B 5:941
9. Artemjeva VN, Smirnova EA, Kukarkin EN, Mikhailova NV, Lyubimova GV, Kudryavtsev VV, Koton MM (1989) Bull Acad Sci USSR, Div Chem Sci 2259

10. Artemjeva VN, Boyarchuk Yu M, Smirnova EA,Kukarkin EN, Kudryavtsev VV, Koton MM (1990) Bull Acad Sci USSR, Div Chem Sci 460
11. Borisova TI, Nikonorova NA, Artemjeva VN, Smirnova EA, Kukarkin EN, Kudryavtsev VV, Koton MM (1990) Bull Acad Sci USSR, Div Chem Sci 463
12. Kudryavtsev VV, Koton MM, Artemjeva VN, Boyarchuk Yu M, Chupans PI (1991) In: Abadie MJM, Silion B (eds) Polyimides and other high temperature polymers. Elsevier, Amsterdam, p61
13. Vinogradova SV, Churochkina NA, Vygodskii Ya S, Zhdanova GV, Korshak VV (1971) Polym Sci USSR, A13:1290
14. Korshak VV, Tsvankin D Ya, Babchinitser TM, Kazaryan LG, Genin Ya V, Vygodskii Ya S, Churochkina NA, Vinogradova SV (1976) Polym Sci USSR, A18:46
15. Azriel A Ye, Gomoreva ZI, Kozaryan LG, Lurye Ye G, Pinayeva NK, Chernova AG (1979) Polym Sci USSR, A18:1496
16. Korshak VV, Babchinister TM, Kazaryan LG, Vasilyev VA, Genin Ya V, Azriel A Ye, Vygodsky Ya S, Churochkina NA, Vinogradova SV, Tsvankin D Ya (1980) J Polym Sci, Polym Phys Ed 18:247
17. Evans JR, Orwoll RA, Tang SS (1985) J Polym Sci, Polym Chem Ed 23:971
18. Koning C, Teuwen L, Meijer EW, Moonen J (1994) Polymer 35:4889
19. Imai Y (1994) Am Chem Soc, Polym Prepr 35(1):399
20. Itoya K, Kumagai Y, Kakimoto M, Imai Y (1992) Polym Prepr Jpn 41:2131
21. Inoue T, Kakimoto M, Imai Y, Watanabe J (1995) Macromolecules 28:6368
22. Inoue T, Kakimoto M, Imai Y, Watanabe J (1997) Macromol Chem Phys 198:519
23. Kumagai Y, Itoya, K, Kakimoto M, Imai Y (1995) Polymer 36:2827
24. Inoue T, Kumagai Y, Kakimoto M, Imai Y (1997) Macromolecules 30:1921
25. Itoya K, Kumagai Y, Kakimoto M, Imai Y (1994) Macromolecules 27:4101
26. Goyal M, Inoue T, Kakimoto M, Imai Y (1998) J Polym Sci, Part A, Polym Chem 36:39
27. Imai Y, Fueki T, Inoue T, Kakimoto M (1998) J Polym Sci, Part A, Polym Chem 36:1341
28. Imai Y, Nemoto H, Kakimoto M (1996) J Polym Sci, Part A, Polym Chem 34:701
29. Morgan PED, Scott H (1969) J Polym Sci, Polym Lett Ed 7:437
30. Morgan PED, Scott H (1972) J Appl Polym Sci 16:2029
31. Ikawa T, Shimamura K, Yokoyama F, Monobe K, Mori Y, Tanaka Y (1986) Sen-i Gakkaishi 42:T-403
32. Ikawa T, Maeda W, Date S, Shimamura K, Yokoyama F, Monobe K (1988) Sen-i Gakkaishi 44:385
33. Imai Y (1996) High pressure polycondensation. In: Salamone JC (ed) Polymeric materials encyclopedia. CRC Press, Boca Raton, 2990
34. Itoya K, Kakimoto M, Imai Y, Fukunaga O (1992) Polym J 24:979
35. Itoya K, Sawada H, Kakimoto M, Imai Y (1995) Macromolecules 28:2611
36. Imai Y (1996) High pressure polymerization processing. In: Salamone JC (ed) Polymeric materials encyclopedia. CRC Press, Boca Raton, 2994
37. Itoya K, Kumagai Y, Kanamaru M, Sawada H, Kakimoto M, Imai Y, Fukunaga O (1993) Polym J 25:883
38. Itoya K, Kakimoto M, Imai Y (1994) Polymer 35:1203
39. Itoya K, Kakimoto M, Imai Y (1994) Macromolecules 27:7231
40. Itoya K, Kakimoto M, Imai Y (1991) Polym Prepr Jpn 40:1821
41. Kricheldorf HR, Pakull R (1987) Polymer 28:1772
42. Kricheldorf HR, Pakull R (1988) Mol Cryst Liq Cryst 157:13
43. Kricheldorf HR, Pakull R (1988) Macromolecules 21:551
44. Kricheldorf HR, Pakull R, Buchner S (1988) Macromolecules 21:1929
45. Kricheldorf HR, Pakull R, Buchner S (1989) J Polym Sci, Part A, Polym Chem 27:431
46. Kricheldorf HR, Schwarz G (1990) Makromol Chem 191:537
47. Kricheldorf HR, Huner R (1990) Makromol Chem, Rapid Commun 11:211
48. Kricheldorf HR, Domschke A, Schwarz G (1991) Macromolecules 24:1011
49. Kricheldorf HR, Pakull R, Schwarz G (1993) Makromol Chem 194:1209

50. Kricheldorf HR, Schwarz G, Berghahn M (1994) Macromolecules 27:2540
51. Pardey R, Zhang A, Gabori PA, Harris FW, Cheng SZS, Adduchi J, Facinelli JV, Lenz RW (1992) Macromolecules 25:5060
52. Pardey R, Sheng D, Gabori PA, Harris FW, Cheng SZS, Adduchi J, Facinelli JV, Lenz RW (1993) Macromolecules 26:3687
53. Pardey R, Wu SS, Chen J, Harris FW, Cheng SZS, Keller A, Adduchi J, Facinelli JV, Lenz RW (1994) Macromolecules 27:5784
54. Sato M, Hirata T, Mukaida K (1992) Makromol Chem 193:1729
55. Hirata T, Sato M, Mukaida K (1993) Makromol Chem 194:2861
56. Hirata T, Sato M, Mukaida K (1994) Macromol Chem Phys 195:1611
57. Asanuma T, Oikawa H, Ookawa Y, Yamashita W, Matsuo M, Yamaguchi A (1994) J Polym Sci, Part A, Polym Chem 32:2111
58. Imai Y (1995) Am Chem Soc, Polym Prepr 36(1):711
59. Imai Y (1996) A new facile and rapid synthesis of polyamides and polyimides by microwave-assisted polycondensation. In: Hedrick JL, Labadie JW (eds) Step-growth polymers for high-performance materials. Am Chem Soc Washington, p421
60. Watanabe S, Hayama K, Park KH, Kakimoto M, Imai Y (1993) Makromol Chem, Rapid Commun 14:481
61. Imai Y, Nemoto H, Kakimoto M (1996) Polym J 28:256
62. Imai Y, Fueki T, Inoue T, Kakimoto M (1998) J Polym Sci, Part A, Polym Chem 36:1031
63. Gaw K, Suzuki H, Kakimoto M (1995) J Photopolym Sci Tech 8:144
64. Gaw K, Suzuki H, Jikei M, Kakimoto M, Imai Y (1996) Mat Res Soc Symp Proc 435:165
65. Morikawa A, Iyoku Y, Kakimoto M, Imai Y (1992) Polym J 24:107
66. Morikawa A, Iyoku Y, Kakimoto M, Imai Y (1992) J Mater Chem 2:679
67. Morikawa A, Yamaguchi H, Kakimoto M, Imai Y (1994) Chem Mater 6:913
68. Iyoku Y, Kakimoto M, Imai Y (1994) Trans Mat Res Soc Jpn 16B:755
69. Iyoku Y, Kakimoto M, Imai Y (1994) High Perform Polym 6:43
70. Iyoku Y, Kakimoto M, Imai Y (1994) High Perform Polym 6:53
71. Iyoku Y, Kakimoto M, Imai Y (1994) High Perform Polym 6:95

Received: March 1998

Processable Aromatic Polyimides

J. de Abajo and J. G. de la Campa

Instituto de Ciencia y Tecnología de Polímeros. CSIC, Juan de la Cierva, 3. 28006 Madrid, Spain (E-mail: ictag35@fresno.csic.es)

This review article deals with aromatic polyimides that are processable from the melt or soluble in organic solvents. Conventional aromatic polyimides represent the most important family of heat resistant polymers, but they cannot be processed in the melt, and their application in the state of soluble intermediates always involves a hazardous step of cyclodehydration and elimination of a non-volatile polar solvent. A major effort has therefore been devoted to the development of novel soluble and/or melt-processable aromatic polyimides that can be applied in the state of full imidation. The structural factors conducive to better solubility and tractability are discussed, and representative examples of monomers showing favourable structural elements have been gathered and listed with the chemical criteria. Experimental and commercial aromatic polyimides are studied and evaluated by their solubility, transition temperatures and thermal resistance. An example is also given of the methods of computational chemistry applied to the study and design of polyimides with improved processability.

Keywords. Polyimides, Aromatic, Monomers, Solubility, Processability, Computer simulation

Advances in Polymer Science, Vol.140

1 Introduction

Aromatic polyimides are considered to be a class of high-performance polymers, and they have found a wide range of applications in advanced technologies [1]. They were developed in the 1960s thanks to intensive research on heat-resistant polymers, and soon became highly important because of their excellent thermal stability, along with good mechanical and electrical properties. On the other hand, it was also soon realized that fabrication of aromatic polyimides was not possible from the melt. For injection or extrusion moulding, conventional aromatic polyimides did not show suitable flow properties, and therefore special fabrication methods such as compression or sintering moulding must be applied. Furthermore, their extreme structural rigidity made them insoluble in any organic media.

Fortunately, aromatic polyimides could be used as materials because they can be prepared through a multistep process, being applicable in the state of soluble polymeric intermediate. Nevertheless, the transformation into polyimides at the moment of application is an approach far from being optimal in most cases, and it can be said that, for many years, aromatic homopolyimides could be successfully applied only in the form of films or coatings [2, 3].

Structural modifications were envisioned early to overcome these limitations. A first improvement was outlined by preparing copolymers, which were soluble in the state of full imidation, mainly poly(ester-imide)s and poly(amide-imide)s [2, 4, 5]. As an alternative to these conventional copolymers, addition polyimides were developed in the 1970s as a new class of thermosetting materials. Thus, bismaleimides, bisnadimides, and end-capped thermocurable polyimides were successfully developed and marketed [6, 7]. These resins were the precursors of the modern PMR (polymeric monomer reactants) formulations [8].

Further improvements in the chemistry of polyimides during the last few years have been directed towards novel, linear species that are soluble in organic solvents or melt-processable while fully imidized. Thus, changes had to be introduced in the chemical structure to adapt the behaviour and performance of these specialty polymers to the demands of the new technologies. As a consequence, a new generation of condensation polyimides has appeared, the so-called thermoplastic polyimides. A recent review on organic-soluble and/or melt-processable thermoplastic polyimides is available [9]. The very high transition temperatures and melt viscosities exhibited by most of the modern aromatic polyimides do not really permit their processing in the melt by the conventional methods of extrusion, compression and injection moulding. Therefore, under the title processable aromatic polyimides, melt-processable and soluble materials have been included in the present revision. Blending with thermoplastics is another approach to make polyimides processable, but this will not be considered here.

Since the commercialization of Kapton polyimide film by DuPont, more than 30 years ago, a great number of polyimides have been described covering a very

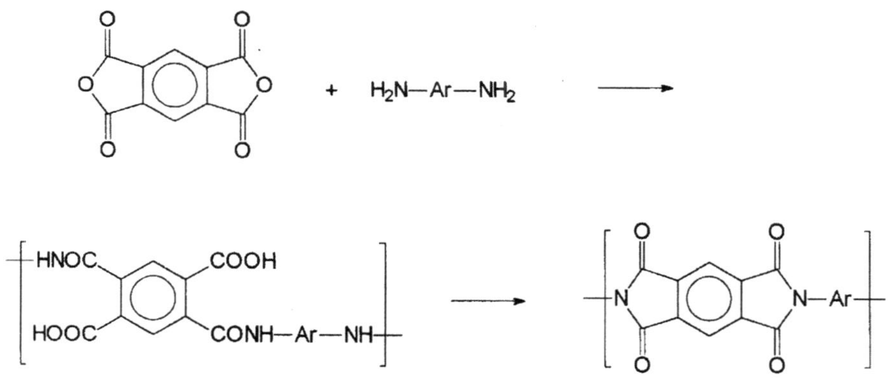

Scheme 1

wide range of properties. However, the properties of Kapton, the polyimide made from pyromellitic dianhydride and 4,4'-oxydianiline, are actually unique, and no other commercial polyimide has challenged the properties-price balance that this high-temperature film offers. Thus, all the efforts dedicated to design and to produce novel aromatic polyimides have concentrated on the processability and on the improvement of some specific properties that are crucial in advanced technologies, such as photosensitivity, transparency, lack of colour, light transmission, dimensional stability, electrical conductivity, moisture absorption, planarization, adhesion, chemical resistance or etching capabilities.

As mentioned before, the first generation of fully aromatic homopolyimides, could be used in a few applications because they had to be applied in the form of soluble polyamic acids, and this limited the materials to be transformed almost exclusively into films or coatings [2, 10]. They all had to be synthesized by a two-step method, as exemplified for an aromatic polyimide from pyromellitic dianhydride in Scheme 1. The method involves the synthesis of a soluble polyamic acid, which, after shaping, can be converted to the related polyimide by a thermal or a chemical treatment. Abundant literature is available on the methods and the mechanisms involved in the synthesis of these polymers [3, 4, 11–13].

The difficulties in processing conventional aromatic polyimides are due to the inherent molecular features of aromatic polyimides, which is particularly true for the most popular of them: polypyromellitimides. Molecular stiffness, high polarity and high intermolecular association forces (high density of cohesive energy) make these polymers virtually insoluble in any organic medium, and shift up the transition temperatures to well above the decomposition temperatures. Thus, the strategies to novel processable aromatic polyimides have focussed on chemical modifications, mainly by preparing new monomers, that provide less molecular order, torsional mobility and lower intermolecular bonding.

Of the various alternatives to design novel processable polyimides, some general approaches have been universally adopted:

- introduction of aliphatic or another kind of flexible segments, which reduce chain stiffness;
- introduction of bulky side substituents, which help for separation of polymer chains and hinder molecular packing and crystallization;
- use of enlarged monomers containing angular bonds, which suppress coplanar structures;
- use of 1,3-substituted instead of 1,4-substituted monomers, and/or asymmetric monomers, which lower regularity and molecular ordering;
- preparation of co-polyimides from two or more dianhydrides or diamines.

However, factors leading to better solubility or lower Tg or Tm in a polymer often conflict with other important requirements, such as mechanical properties, thermal resistance or chemical resistance. Therefore, an adjusted degree of modification should be applied to optimize the balance of properties.

The computer calculation of some conformational parameters, such as bond angles, rotation barriers, and minimal energies of conformational isomers, can nowadays give valuable information to explain the behaviour of novel materials, and to predict the properties of polymers synthesized from new monomers. An illustrative example of these theoretical methods, is given in the appendix.

2 Polyimides Containing Aliphatic and Other Flexible Spacers

Aliphatic-aromatic polyimides were discovered early at the start of the polyimides era. They can be prepared from aromatic dianhydrides and aliphatic diamines by several methods, in solution or by melt fusion of salts from a diamine and a tetracarboxylic acid or diacid-diester [2, 14, 15]. Aliphatic-aromatic polyimides did not warrant much attention for many years, probably because the first property sought in polyimides has traditionally been thermal resistance, and the presence of aliphatic groups in the main chain, or as side substituents, was quickly associated with better solubility and processability, but also to much poorer stability. In fact, they did not show Tm much higher than conventional nylons, and the main advantage, their high Tg values, did not mean any useful advance as their performances under service conditions were comparable to the semicrystalline aliphatic polyamides, which in turn are much cheaper. Nevertheless, many aliphatic-aromatic polyimides were described, mainly polypyromellitimides, and some of them were melt-processable. For instance, the polyimides from pyromellitic anhydride and long (seven or more methylene linkages) α,ω-alkylenediamines could be processed at 320–330 °C, and mouldable polyimides could be attained from diamines containing as few as four carbon atoms between amino groups if they are combined with non-planar dianhydrides like oxydiphthalic anhydride or 2,2-bis(3,4-dicarboxyphenyl)propane dianhydride [15, 16].

Table 1. Monomers for polymides containing flexibilizing spacers

Monomer	Reference
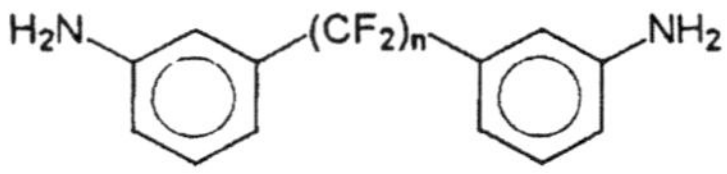	23
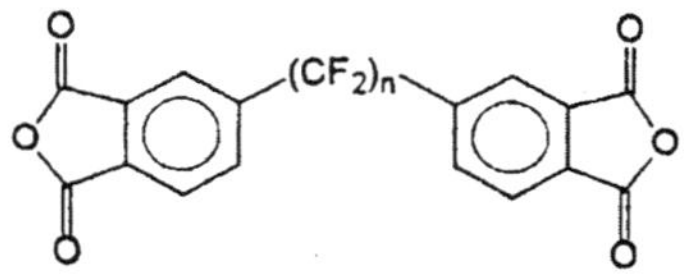	23,30
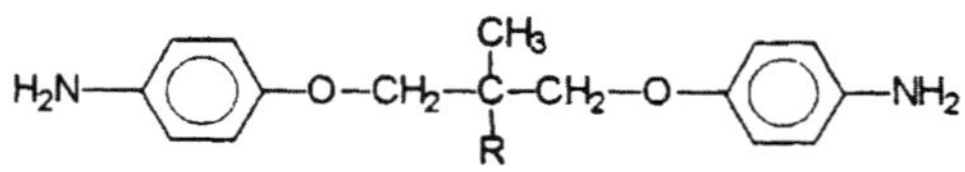	33,34
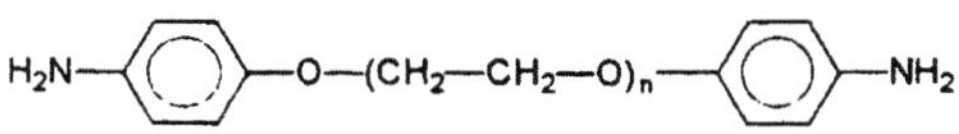	41
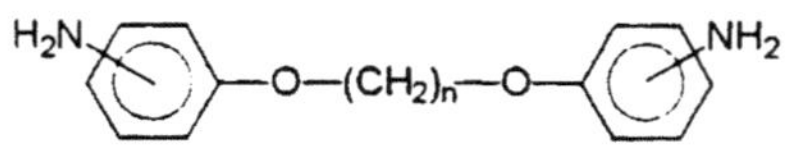	31
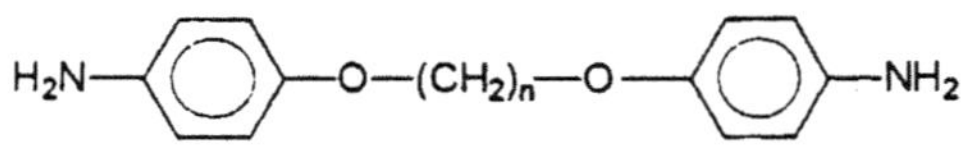	25
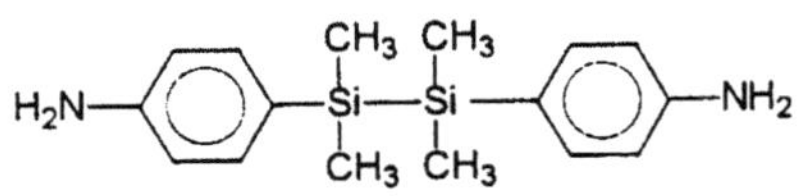	37
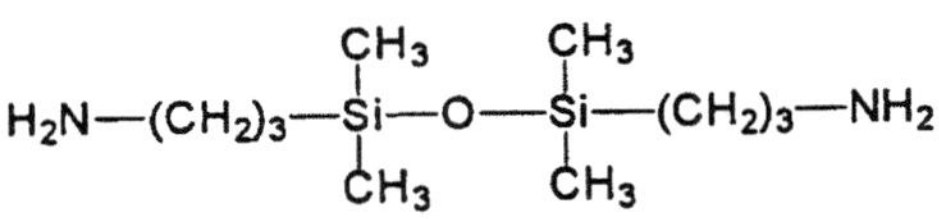	38,39
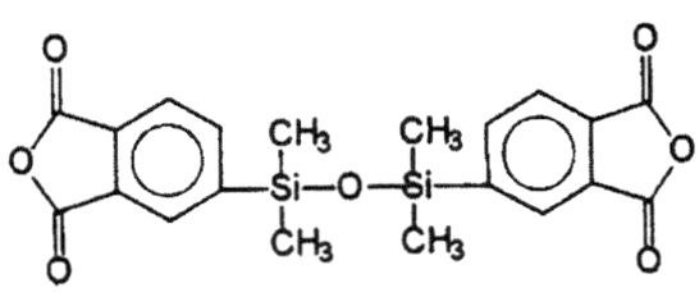	29

Unlike aliphatic homopolyimides, aliphatic-aromatic copolyimides have achieved great importance, and many linear poly(ester-imide)s and poly(amide-imide)s containing aliphatic linkages have been reported, mainly those derived from the trimellitimide unit [17–20]. Although they have not received technical development, they have served to open new alternatives to the application of polyimides as processable materials, and have meant a valuable contribution to the knowledge of the structure-properties relationships that govern the behaviour of these polymers [21, 22]. In the last few years, aliphatic-aromatic LC-polyimides containing ester, amide, or ether groups have received much attention, and they are the subject of another part of this volume.

In recent years polyimides containing flexible spacers, aliphatic or not, have gained importance as technical materials. Thus, some segmented, fluorinated polyimides [23], and co-polyimides containing oligoethylene glycol sequences have been presented as processable polyimides with potential application as thermally stable adhesives and thermoplastics [24, 25]. Polyimides containing oligosiloxane segments can also be included within this class of polyimides [26–28]. They could all really be considered as block copolymers based on aromatic polyimide and short chains of polyethylene glycol, polysiloxane, alkane or fluoroalkane. The thermal stability of these segmented polyimides is dependant on their chemical structure, and mainly on the flexible chain. Thus, polyimides containing polyethylene glycol sequences are thermally unstable [24], whereas polysiloxane [28, 29] and perfluoroalkanes [30], and even alkanes [31, 32], provide a reasonable thermal stability.

For the preparation of these segmented polyimides or block copolyimides, diamines rather than dianhydrides are preferred as the monomer bearing the flexible segment. Some of these monomers have been listed in Table 1.

The thermal stability of one of these polyimides is compared with that of poly(4,4'-oxydiphenylene) pyromellitimide in Fig. 1. Perfluoroalkylenes are the most recommendable groups to be introduced as flexibilizing moieties as they provide a substantial lowering of the Tg without greatly impairing the thermal, mechanical and chemical properties. In spite of this, the thermal stability measured by TGA was significantly higher for the fully aromatic polyimide than for the polyimide having a perfluoroalkylene spacer [30].

Novel soluble and meltable aromatic polyimides containing flexible linkages of the type mentioned above have been described in the last few years. They have been developed as a response to the growing demand of specific materials for advanced technologies, in particular for microelectronics, commercial aircraft applications and aerospace industry. The combination of conventional and new monomers are providing such an expansion of the scope that is possible to confirm a true revival in processable polyimides containing flexible spacers [33–41]. Several new families offer considerable promise in this area, and they have potential use as adhesives, matrix materials for fibre-reinforced composites, or electronic components. All of them are soluble in some specific organic media, showing glass transition temperatures in the range 250–380 °C.

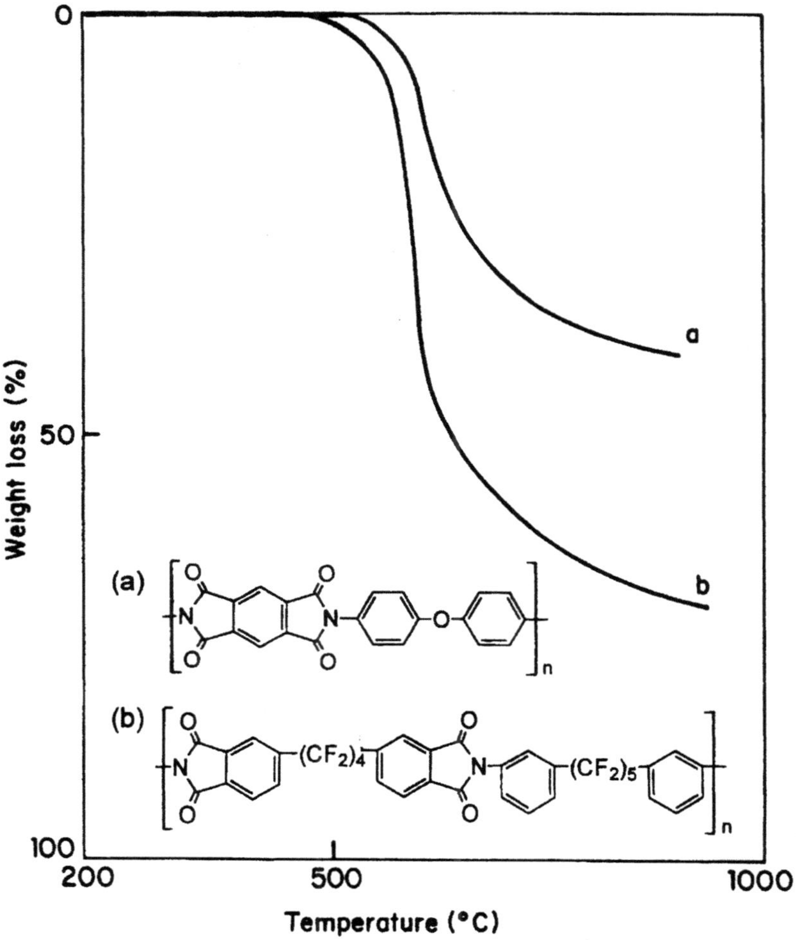

Fig. 1. Comparative TGA curves of poly(4,4'-oxydiphenylene) pyromellitimide and a perfluoro-alkylene polyimide30 (with permission)

3 Polyimides with Bridging Functional Groups

Introducing flexible linkages into polymer backbones is a general approach, used mainly to lower transition temperatures and to improve solubility of intractable aromatic polymers. Polyimides with flexible linkages have been known from the advent of high temperature aromatic polyimides. In fact, most of the commercial, fully aromatic polyimides contain ketone or ether linkages in their repeating units, and early works in the field soon demonstrated that dianhydrides having two phthalic anhydride moieties joined by bonding groups gave more tractable polyimides [42–45].

Many different linkages have been introduced with these purposes, but the most promising are -O-, C=O, -S-, $-SO_2-$, $-C(CH_3)_2-$, $-CH_2-$, -CHOH-, and $-C(CF_3)_2-$. These bonding groups may be located on the dianhydride, on the diamine or on both monomers, or they can even be formed during the polycondensation reaction, when some functional monomers containing preformed phthalimide groups are used as condensation monomers [46, 47]. The presence of flexible linkages has a dramatic effect on the properties of the final polymers. First, "kink" linkages between aromatic rings or between phthalic anhydride functions cause a breakdown of the planarity and an increase of the torsional mobility. Furthermore, the additional bonds mean an enlargement of the repeating unit and, consequently, a separation of the imide rings, whose relative density is actually responsible for the polymer tractability. The suppression of the coplanar structure is maximal when voluminous groups are introduced in the main chain, for instance sulfonyl or hexafluoroisopropylydene groups, or when the monomers are enlarged by more than one flexible linkage.

Some diamines and dianhydrides with a flexible linkage in their structure have been listed in Table 2. The combination of those dianhydrides and diamines, and also the combination of some of them with conventional rigid monomers like benzenediamines, benzidine, pyromellitic dianhydride or biphenyldianhydride, offer a major possibility of different structures with a wide spectrum of properties, particularly concerning solubility and meltability [48–56].

However, very few of the polymers that can be synthesized combining monomers of Table 2, have been reported as melt-processable, although many of them are soluble in organic highly polar solvents. All of them show high glass transition temperatures, commonly over 250 °C, and, theoretically, they can develop crystallinity upon suitable thermal treatment, mainly those containing polar connecting groups. Thus, depending on the nature of X and Y in the general formula, polyimides can be prepared that show an acceptable degree of solubility

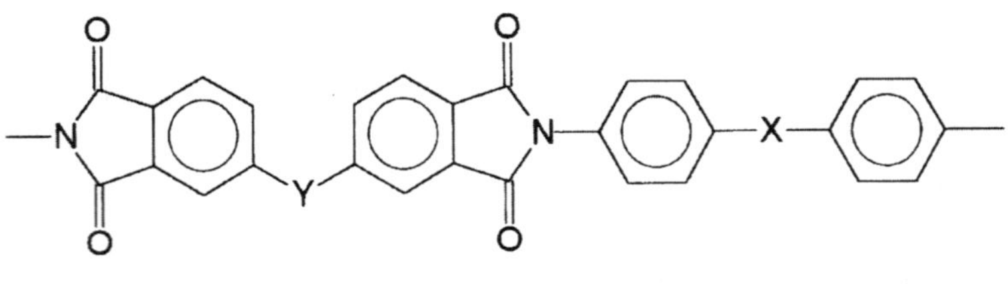

in organic solvents [44, 45]. Hexafluoropropylidene, carbonyl and sulfonyl are the groups most advantageously incorporated concerning processability. This is due to the relatively large volume of these groups, and to the conformational characteristics imparted by them to the polymer chain.

Table 3 shows the Tg values and solubility of some selected polyimides among those prepared from monomers of Tables 1 and 2. The combination of non-planar dianhydrides and non-planar, *meta*-oriented aromatic diamines containing flexible linkages provides the structural elements needed for solubility and melt processability. Some aromatic polyimides marketed as thermoplastic materials are based on these statements [9, 57–60].

Table 2. Diamines and dianhydrides with a flexible linkage

Diamines	Dianhydrides
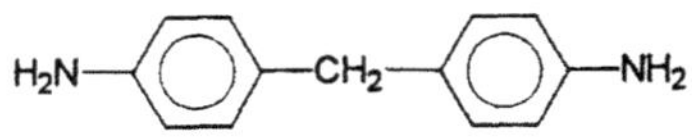	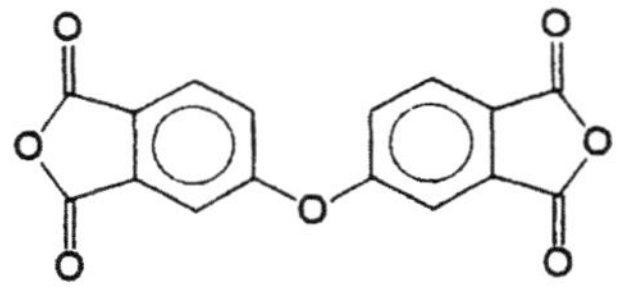
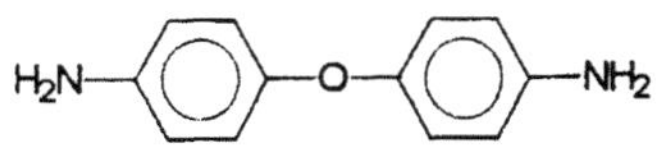	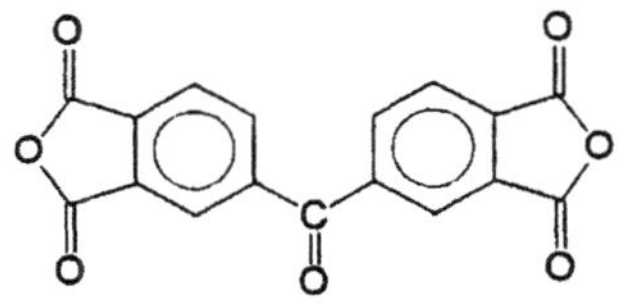
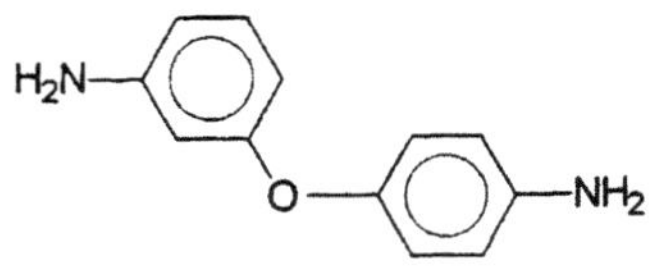	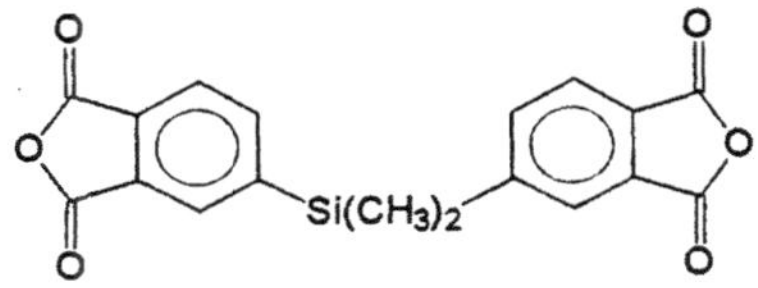
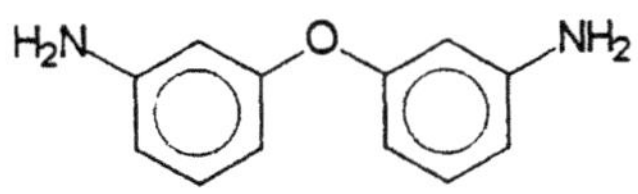	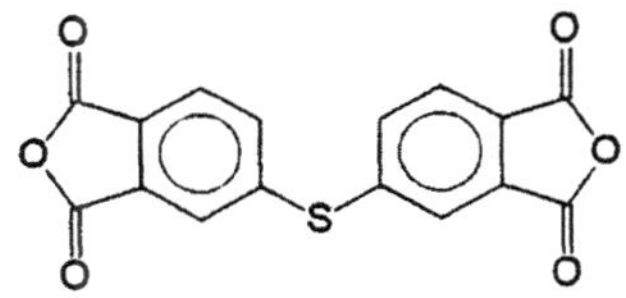
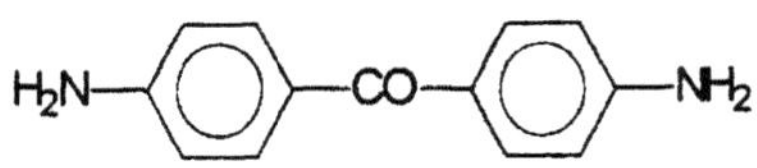	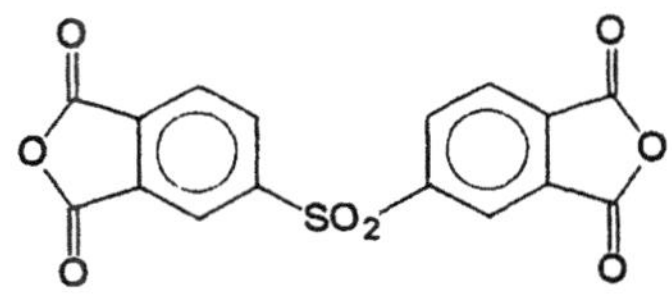
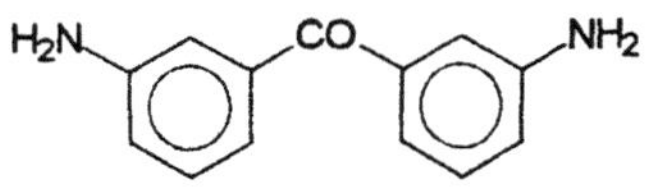	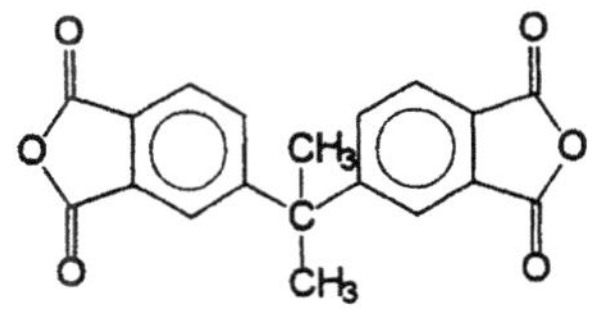
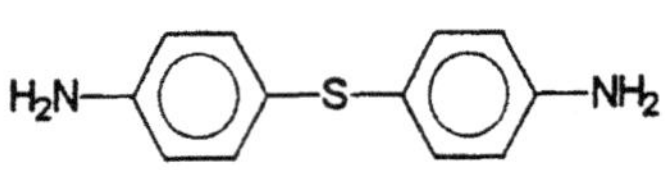	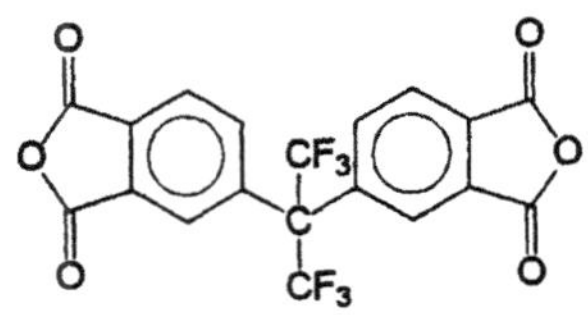

Table 2. (continued)

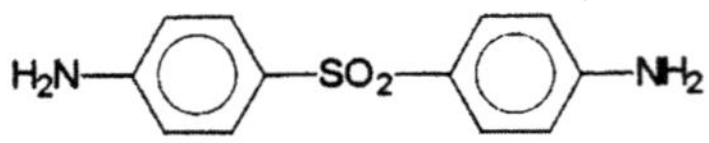

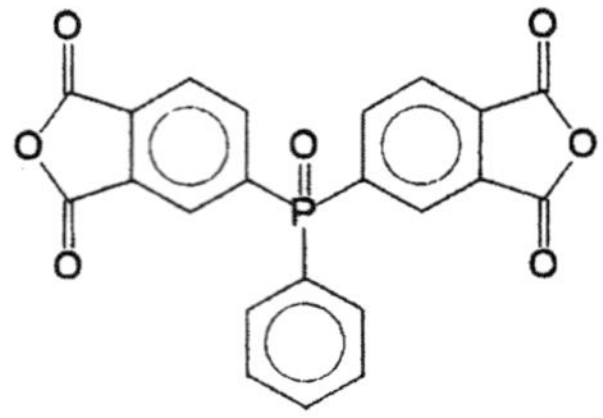

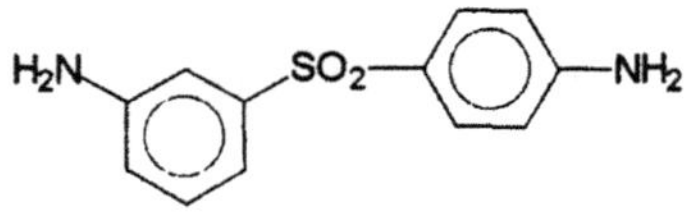

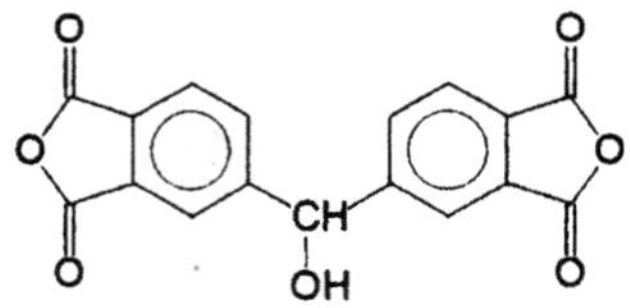

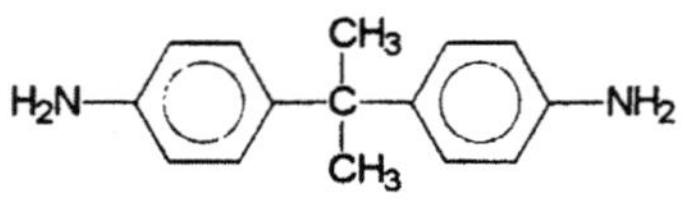

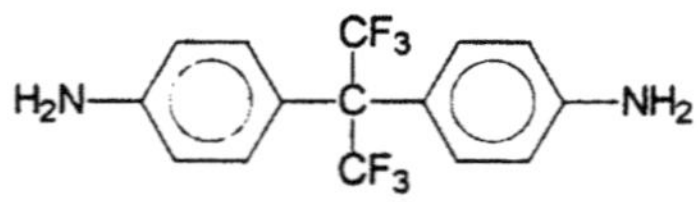

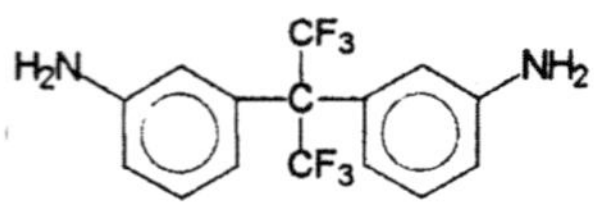

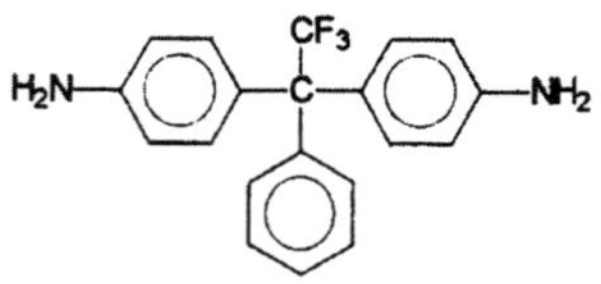

A rational approach to improve tractability of aromatic polyimides consists of extending the length of the diamine and/or the dianhydride by incorporating more phenylene rings and more connecting flexible groups. Monomer enlargement contributes to the separation of the very polar, rigid phthalimide groups, and provides an improvement of the chain mobility by the presence of additional bonds with lower rotational barriers.

Using the same chemical functions previously mentioned for bridged diamines and dianhydrides, a wide variety of new enlarged monomers has been

Table 3. Properties of selected polyimides from monomers with one flexible bridging group

Polymer	Ref.	Tg, °C	Solubility NMP	Solubility *m*-cresol
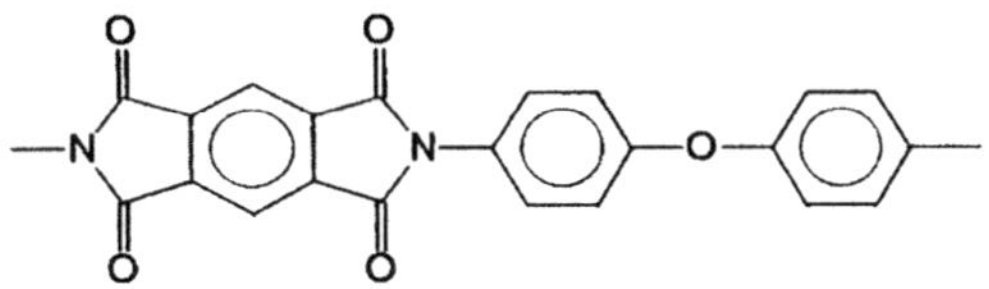	48 49 50	399 370 361	–	–
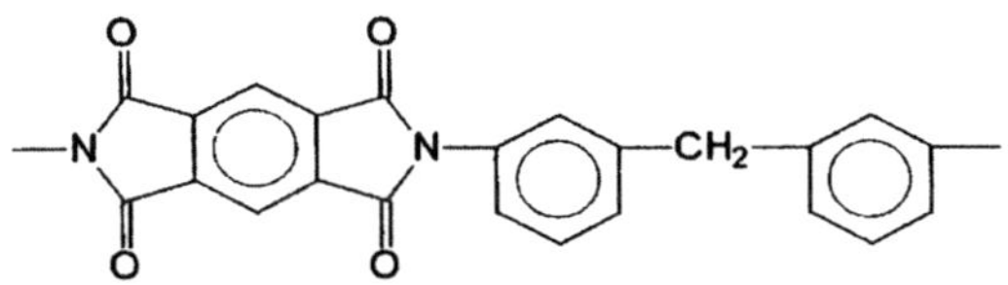	48	308	–	–
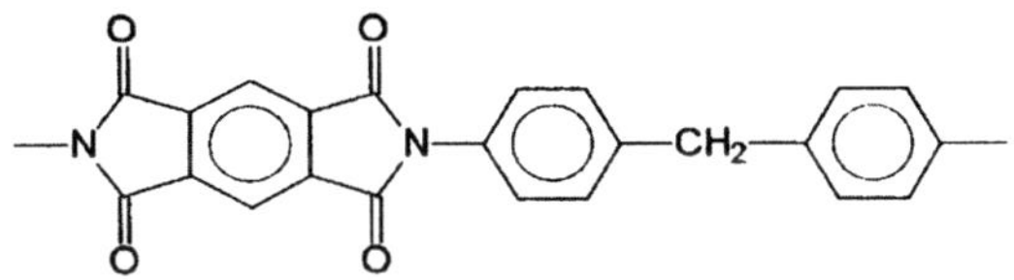	48	339	–	–
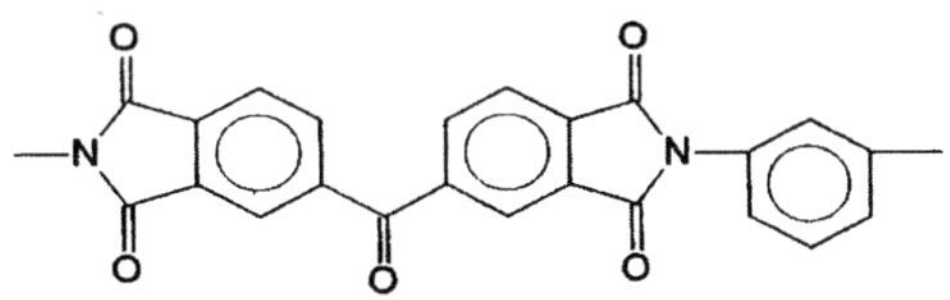	48	300	–	–
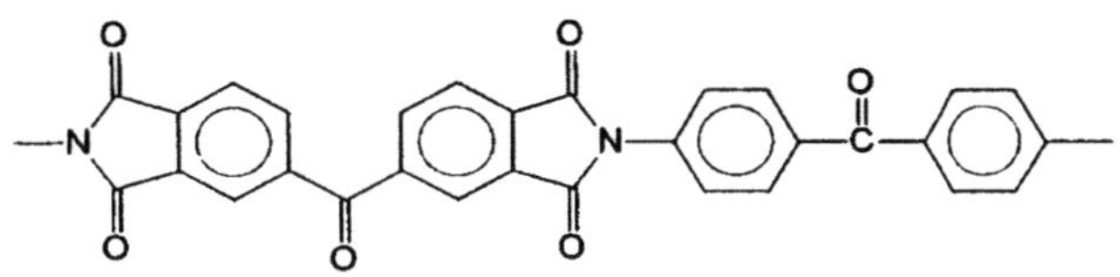	48	288	–	

Table 3. (continued)

Polymer	Ref.	Tg, °C	Solubility NMP	Solubility *m*-cresol
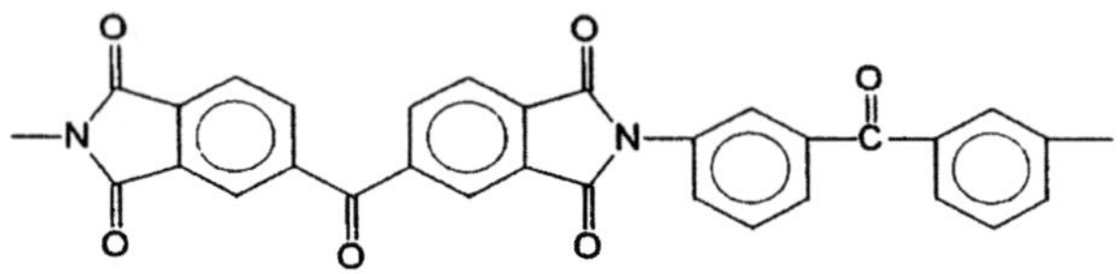	48	259	–	±
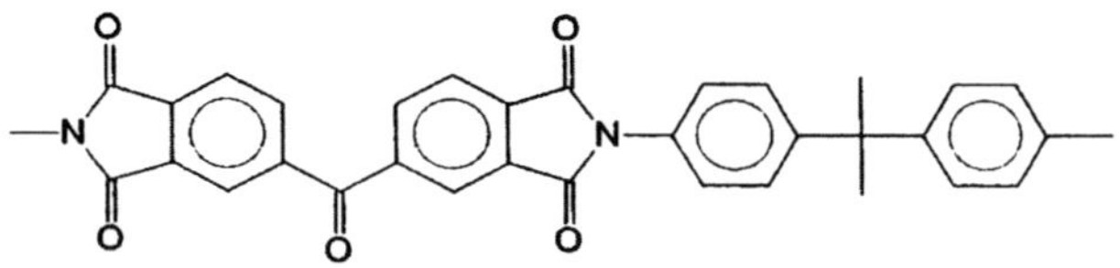	53	297	+	+
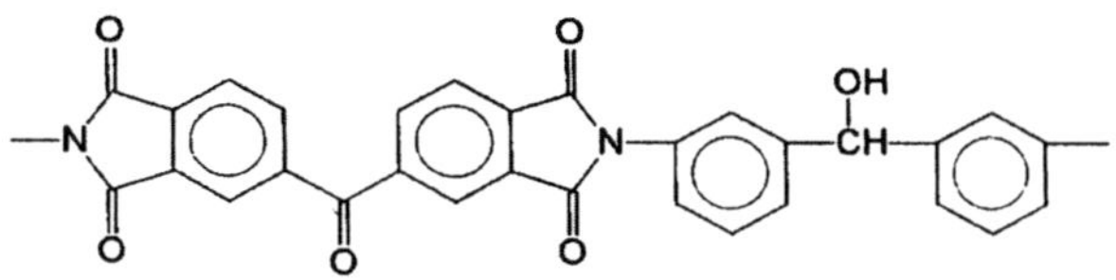 	55	267	++	++
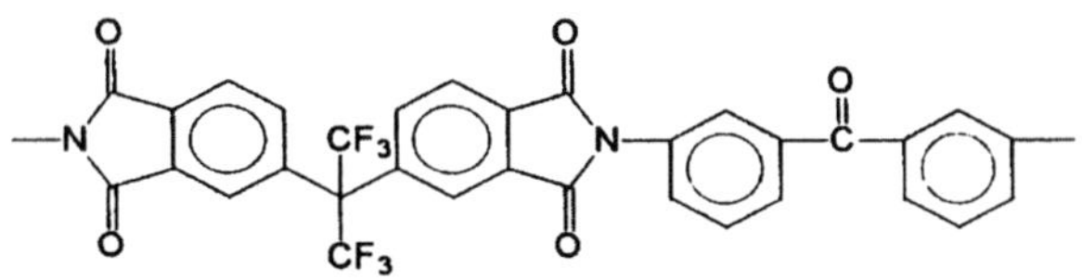 	70	311	+	±
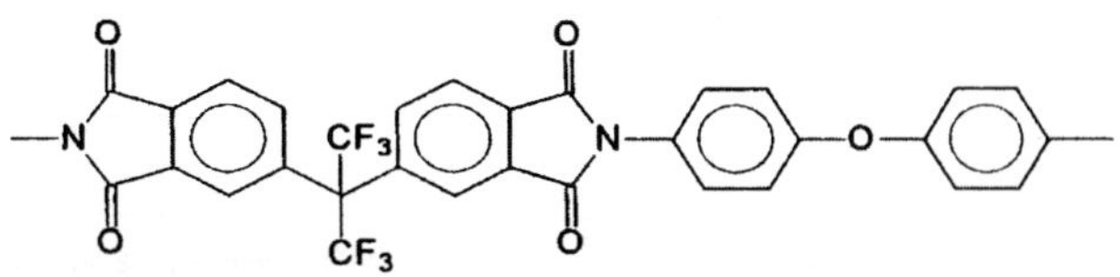 	119 141	299 297	–	+

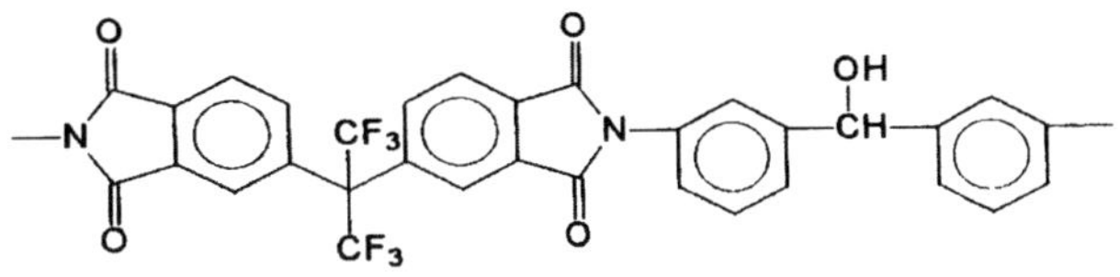	55	262	++	++
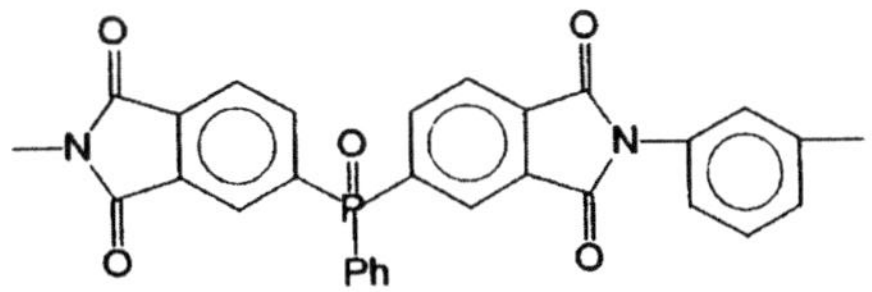	51	323	+	
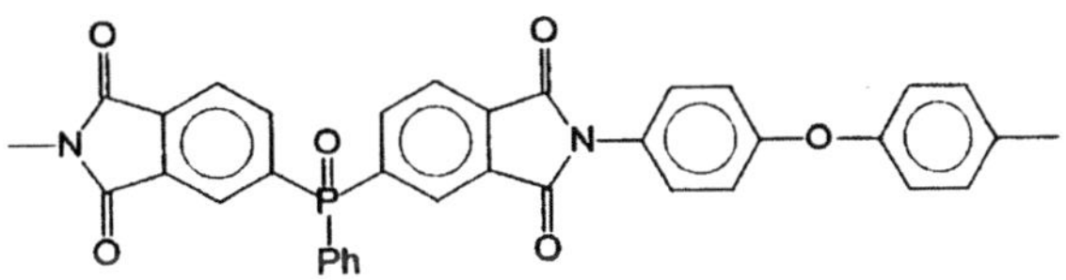	51	318	+	
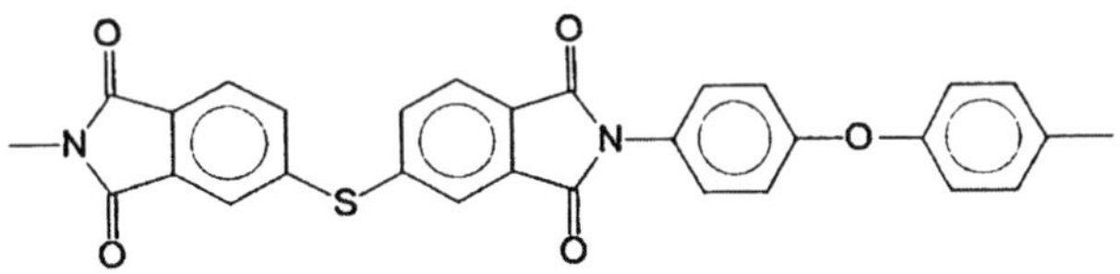	49	270		
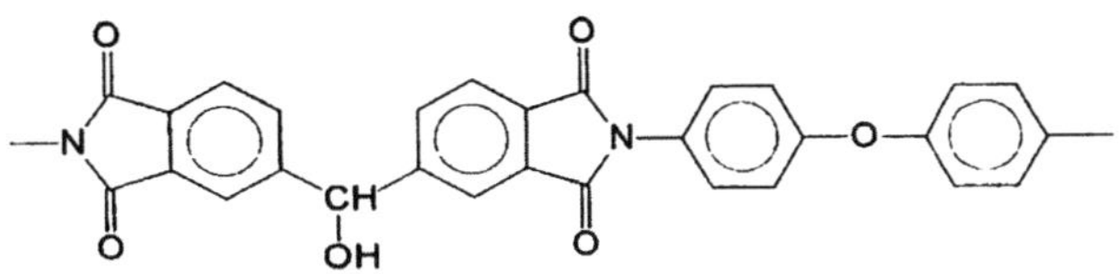	52	284	++	++

Table 3. (continued)

Polymer	Ref.	Tg, °C	Solubility NMP	*m*-cresol
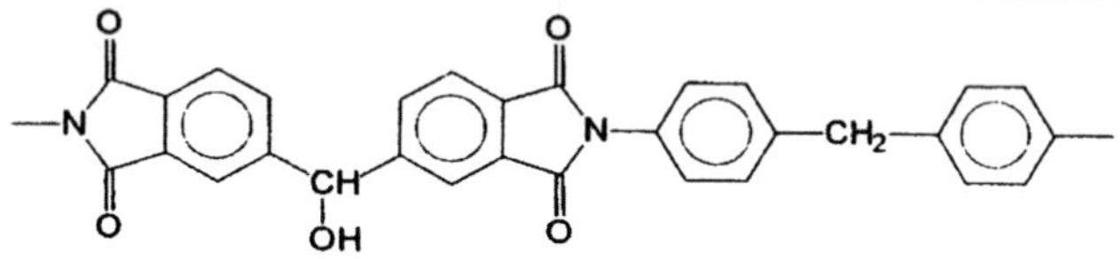	52	296	++	++
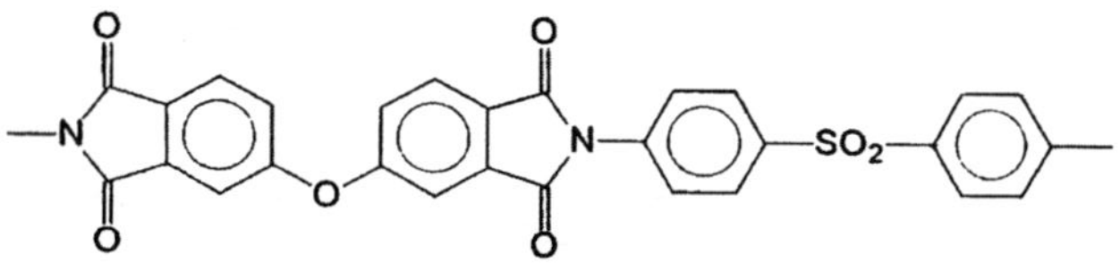	54 42	263 300	++	+
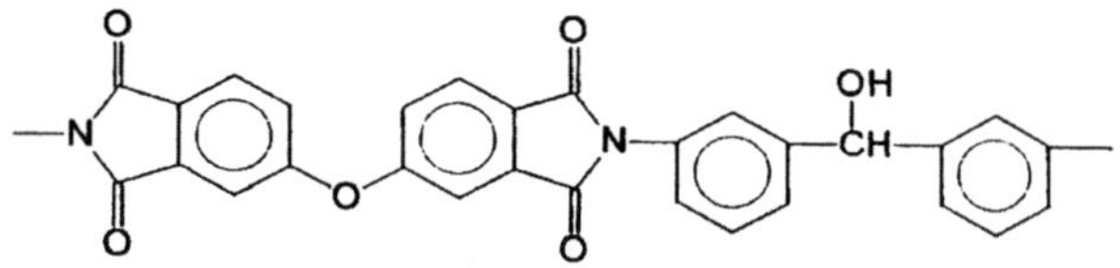	55	252	++	++
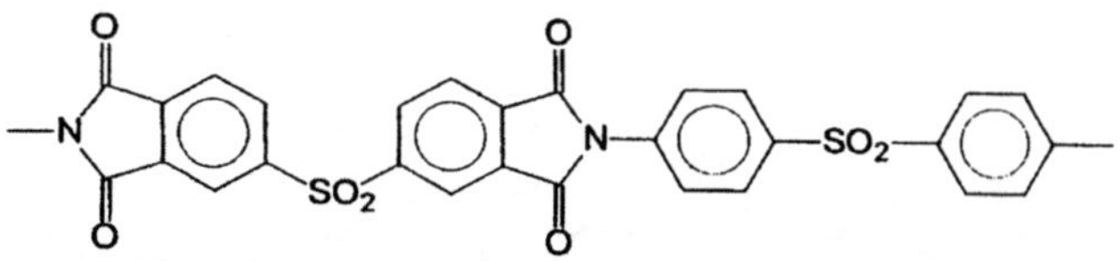	56	307	++	++

++ Soluble at room temperature, + Soluble in hot, ± Partialy soluble or swollen, – Insoluble

Table 4. Monomers for polyimides containing more than one flexible linkage

Diamine	Reference	Dianhydride	Reference
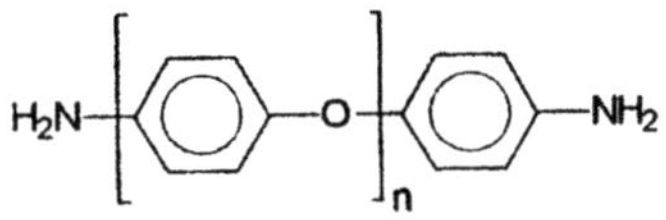	61,62	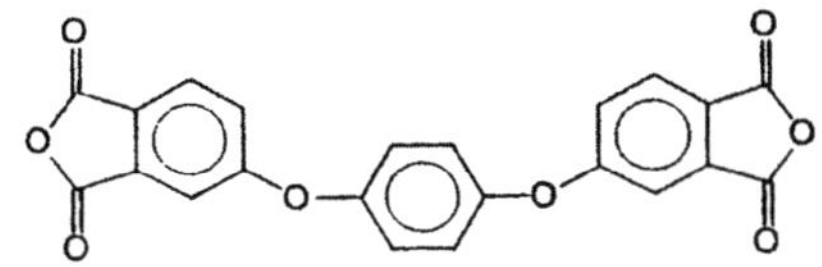	49, 64, 80
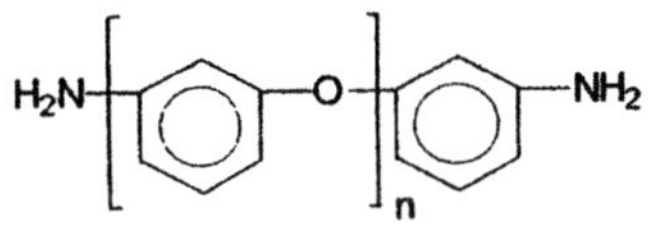	63	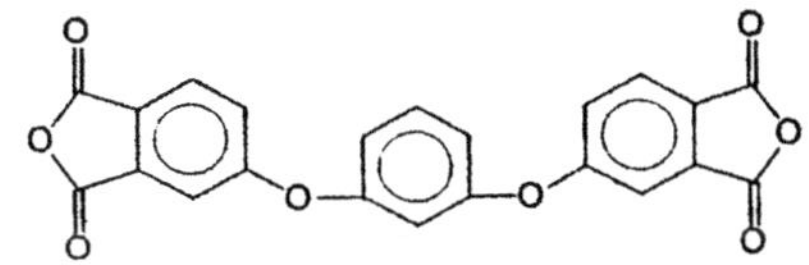	49, 64, 80
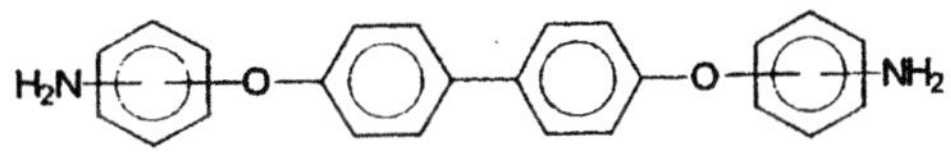	62, 64, 65, 66,	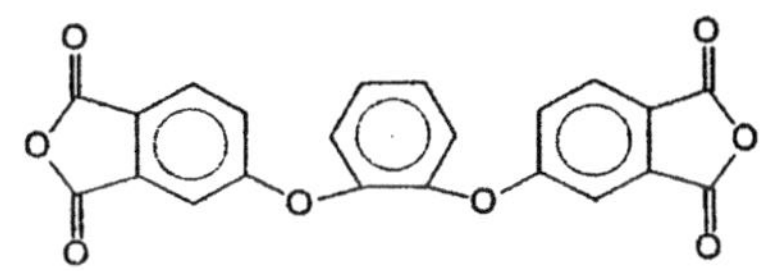	49, 64, 80
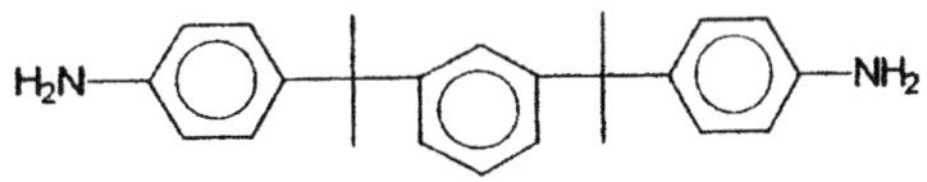	80	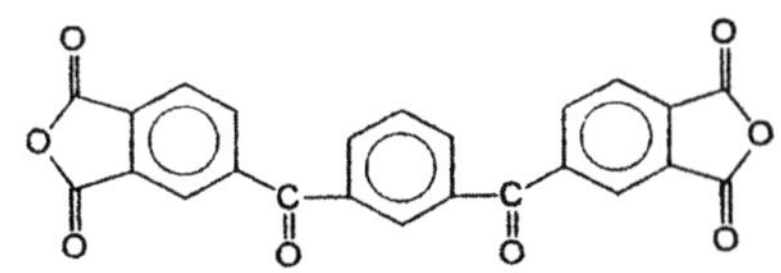	81
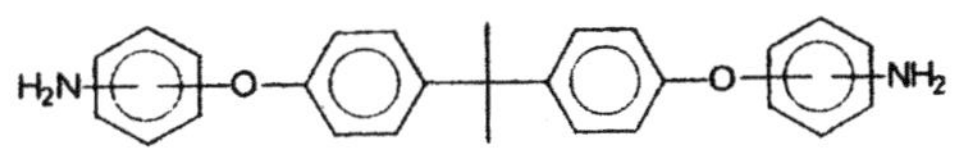	68	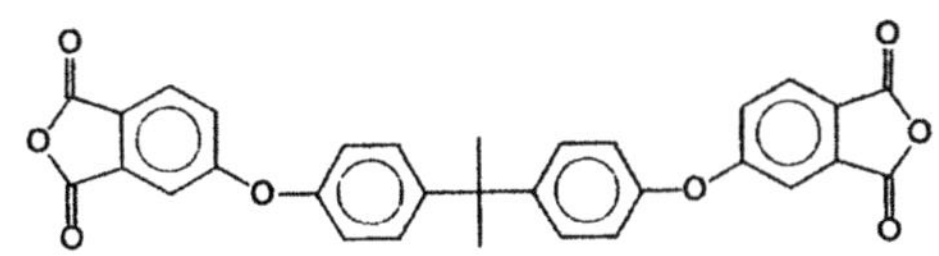	59, 82

Table 4. (continued)

Diamine	Reference	Dianhydride	Reference
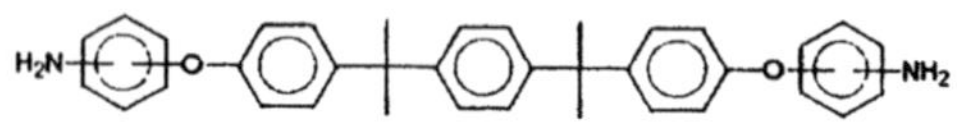	76	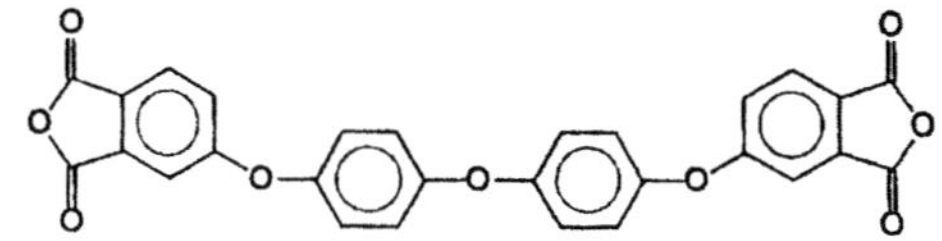	59, 82
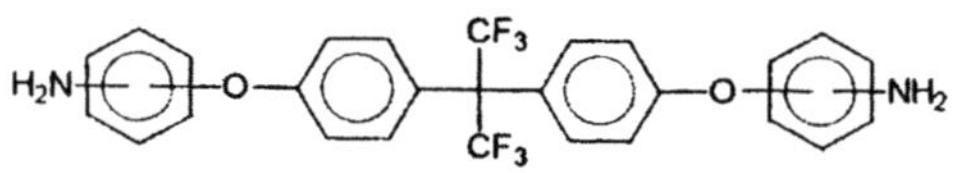	69, 70, 71, 72	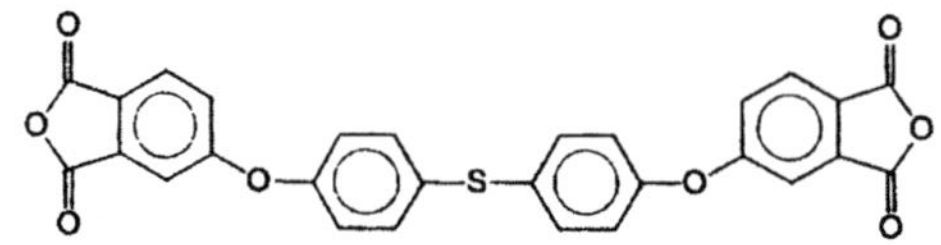	63
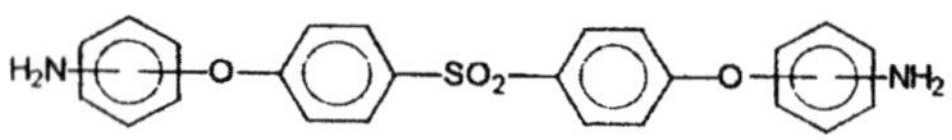	67	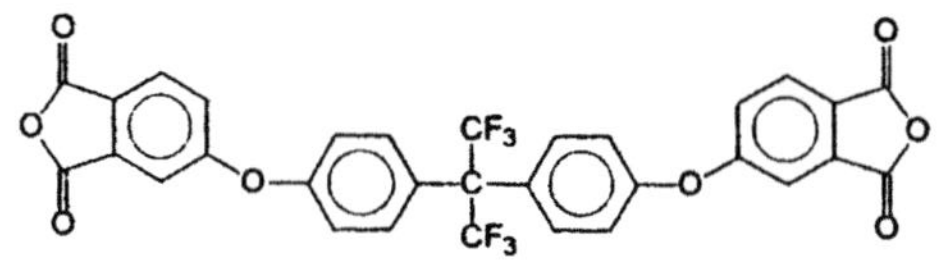	70
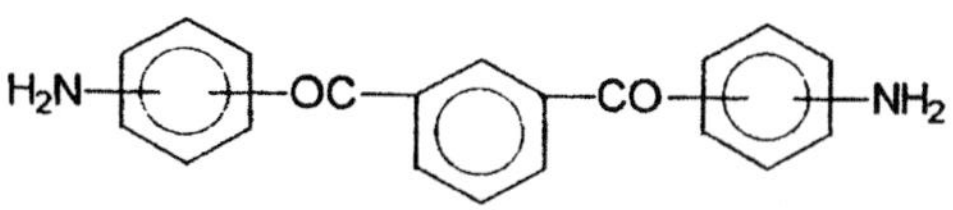	73	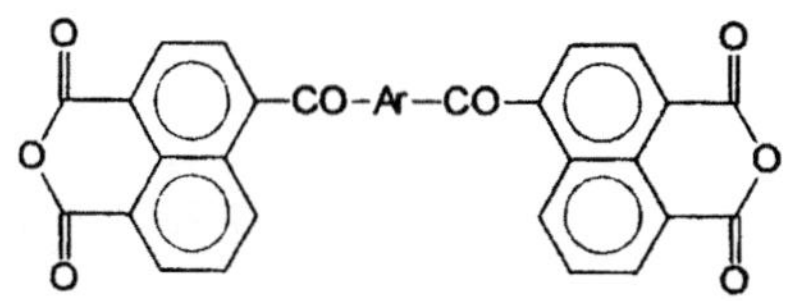	83

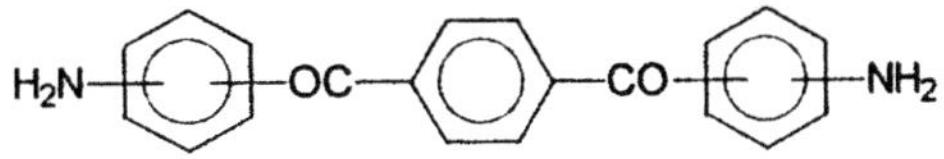	73	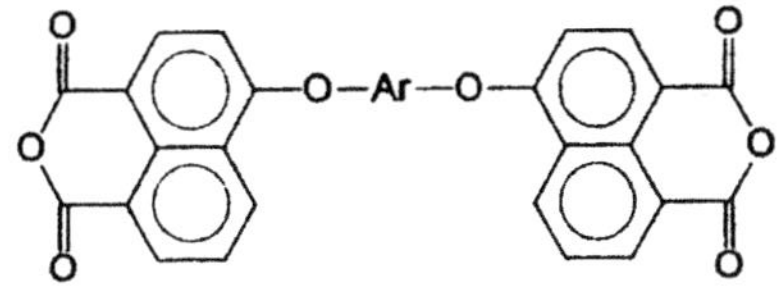	84
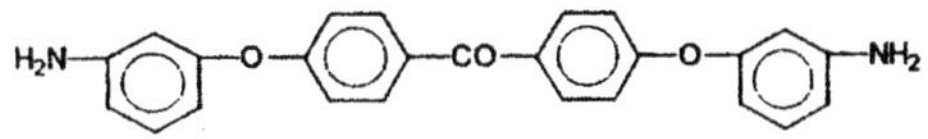	78		
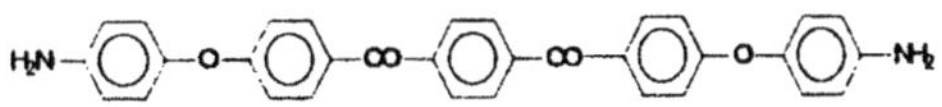	74, 77, 94		
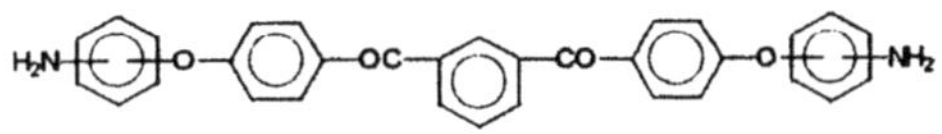	74, 75		
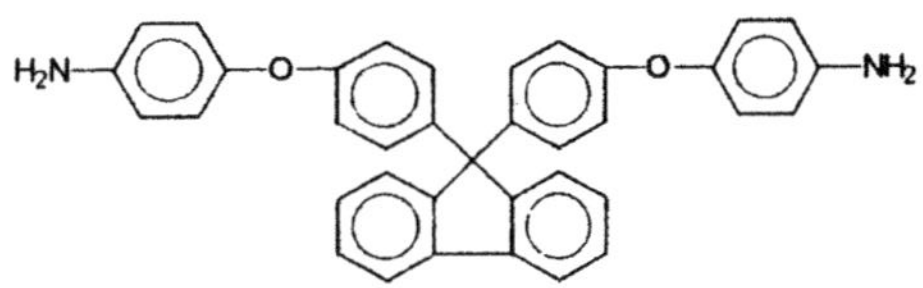	79		

prepared and used to synthesize processable polyimides [61–84]. Some representative monomers, diamines and dianhydrides having more than one bridging group have been listed in Table 4. Although very few of them have achieved commercial importance, the research effort during the last two decades to develop new monomers has enriched the chemistry of aromatic polyimides to such an extent that it would be difficult to find another field of macromolecular chemistry where the investigations have produced a similar variety of new species with such a wide range of properties.

As to the polymers, the most important of them – considering production figures – are very probably the poly(ether-imide)s (PEIs), marketed under the trade name Ultem. Neat PEI resins are amorphous, soluble polymers that show Tg values around 220 °C. They can be processed from the melt by conventional means, and offer a price-performance balance that enables them to compete successfully in the market of engineering thermoplastics.

4
Polyimides with Bulky Side Substituents

Structural modifications to attain soluble aromatic polyimides have also been carried out by introducing bulky substituents, aryl or heterocyclic rings. One of the first references to this approach was made by Korschak and Rusanov, who synthesized soluble aromatic polyimides containing side phthalimide groups [85]. More recent work by Rusanov et al. has enriched this topic with new soluble polyimides containing pendent imide groups [86, 87].

Since the pioneering works of the Russian researchers, many attempts have been made to prepare new monomers, diamines and dianhydrides, with bulky pendent groups for novel processable polyimides. Table 5 shows some of these monomers. Monomers with small side substituents, like methyl, methoxy, trifluoromethyl or halogen, have not been included. In this respect, it is worth noting that many monomers derived from biphenyl, containing small side substituents, have been discovered and used in the synthesis of soluble polyimides in the last few years. They are not going to be discussed in depth in this chapter as they are considered in other parts of this book.

By far, the most promising species are those containing phenyl pendent groups. The phenyl group does not introduce any relevant weakness regarding thermal stability, and provides a measure of molecular irregularity and separation of chains very beneficial in terms of free volume increase and lowering of the cohesive energy density [88–94]. Fluorene diamines and the so-called "cardo" monomers also mean valuable alternatives for the preparation of processable polyimides [95, 96]. On the other hand, the presence of the bulky side substituents in polyimides or in any other linear polymer causes a lowering of the chain's torsional mobility and generally an increment of the glass transition temperature [97–100].

Table 5. Diamines and dianhydrides used in the preparation of polyimides with bulky side groups

Diamine	Reference
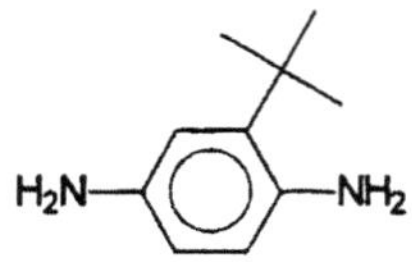	103
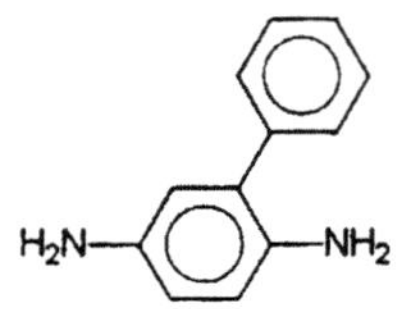	103
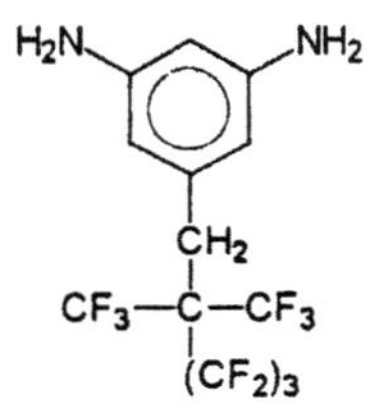	106
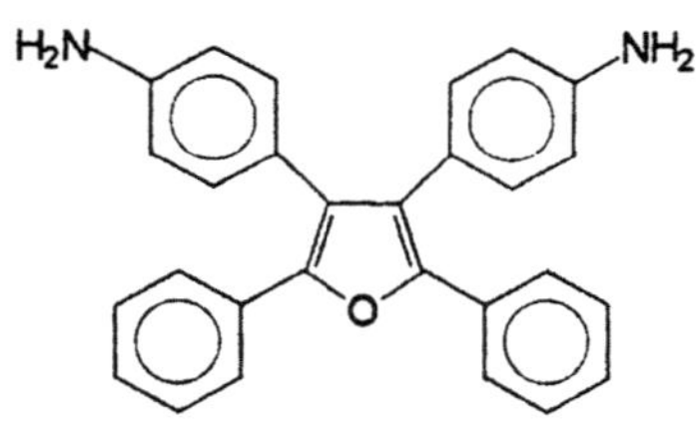	102
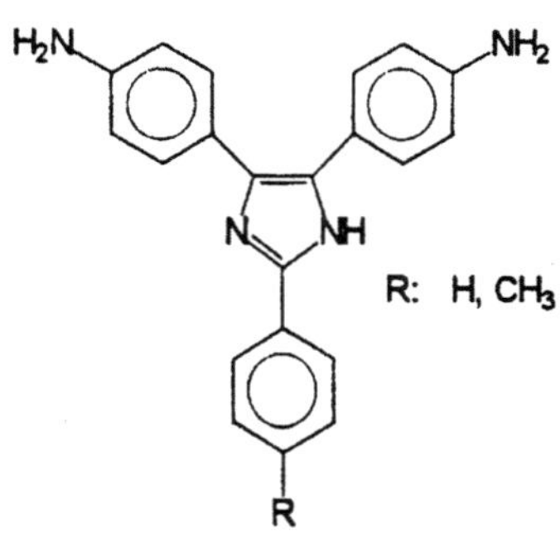	93

Table 5. (continued)

Diamine	Reference
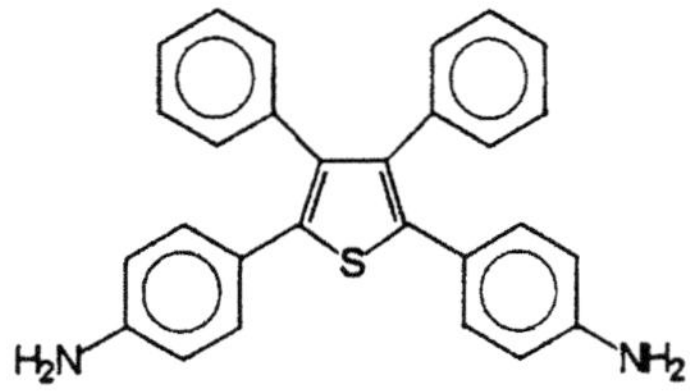	92
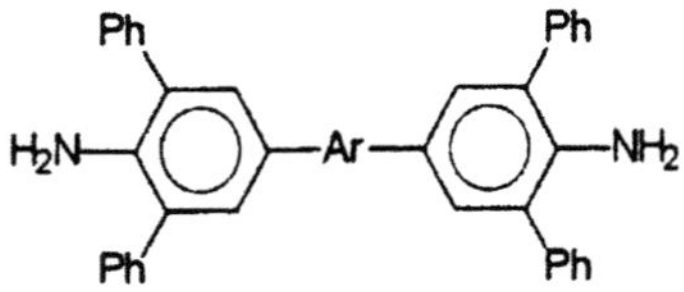	94, 104
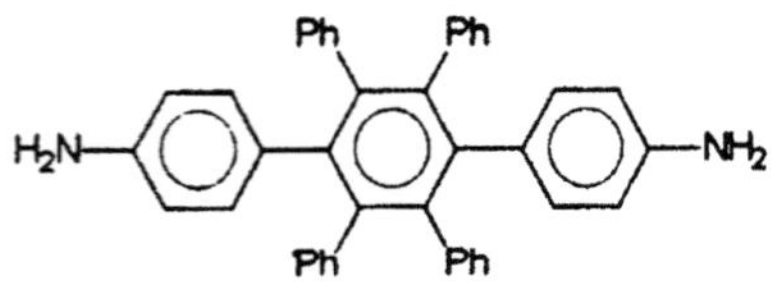	104
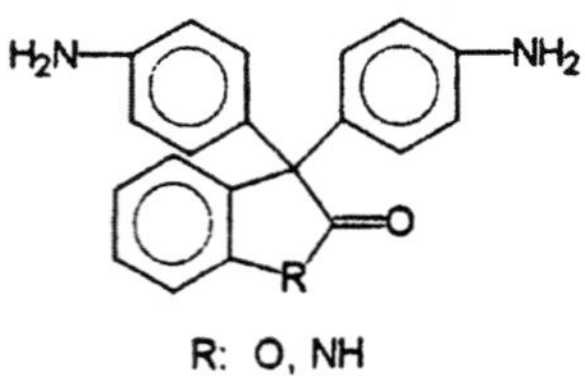	95
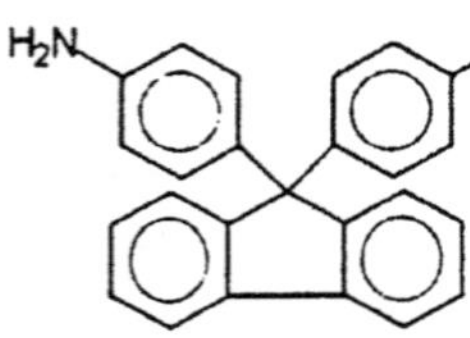	95

Table 5. (continued)

Diamine	Reference
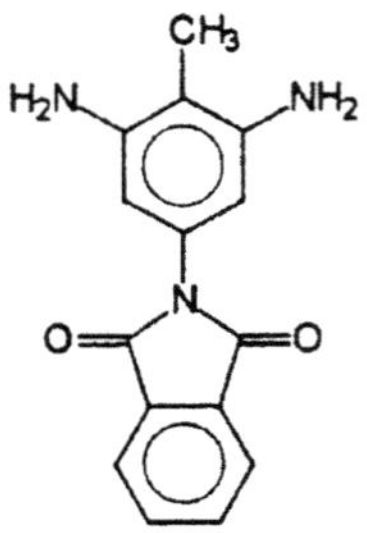	86
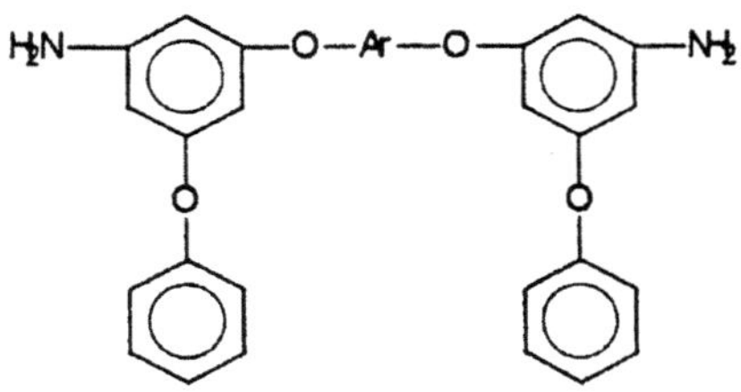	86
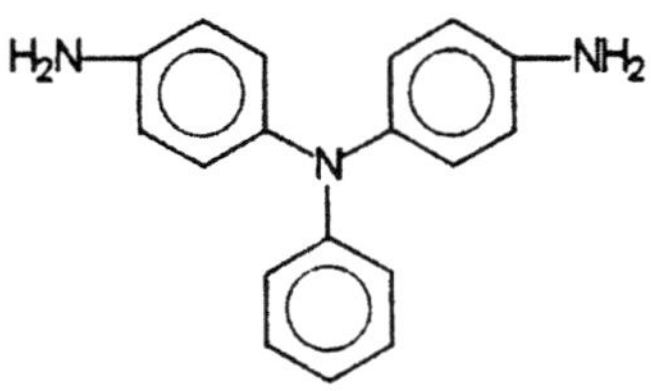	105
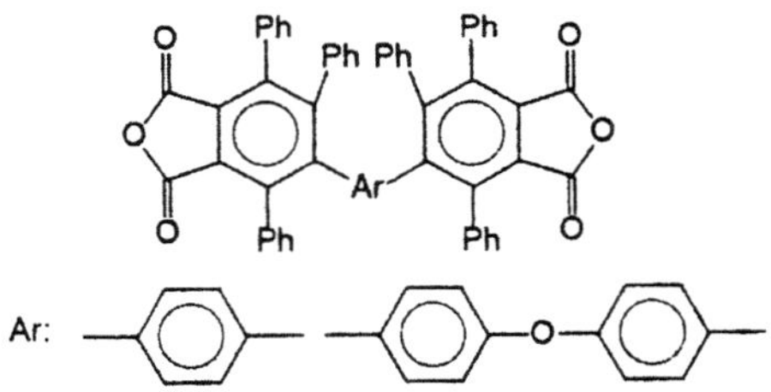	88
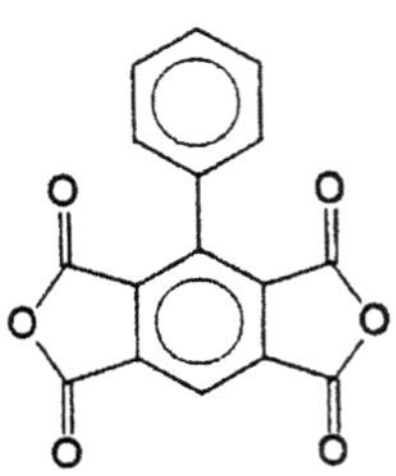	90

Table 5. (continued)

Diamine	Reference
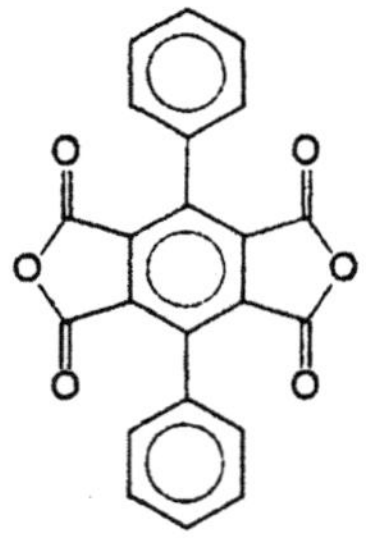	89, 90
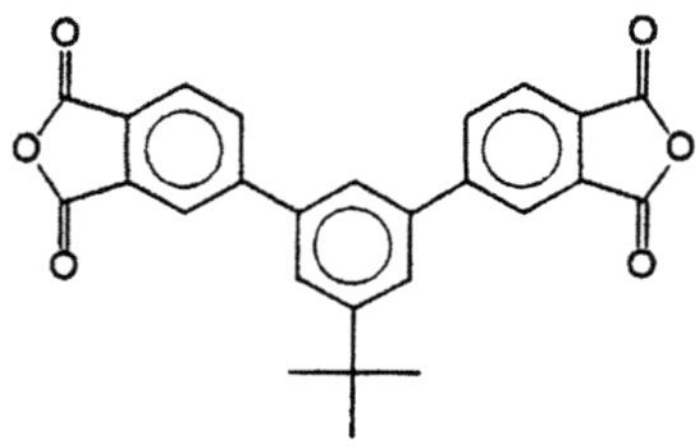	91
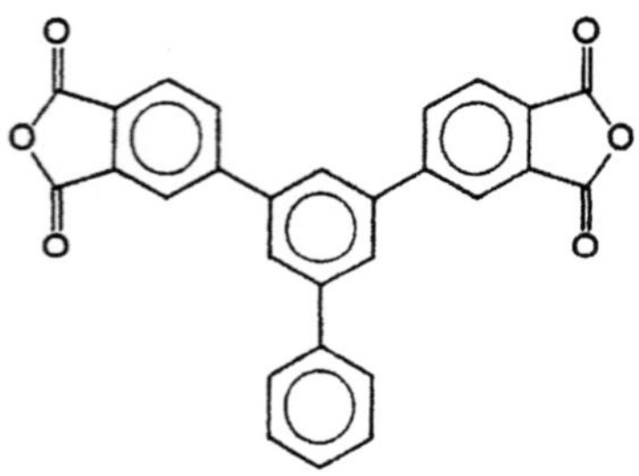	91
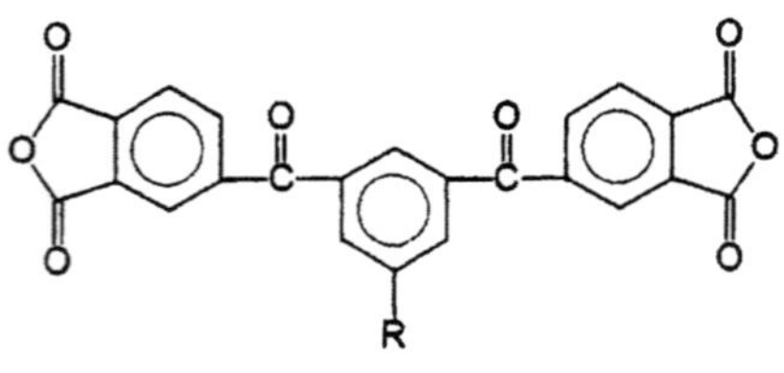 R: *t*-But, Ph, NO_2Ph	101
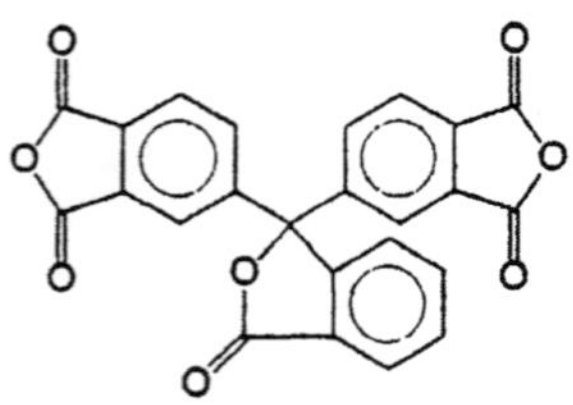	74

5 Other Processable Polyimides

Fully imidized soluble polyimides have ben prepared using monomers derived from diphenylindane and aromatic dianhydrides. Technical polymers (XU218, for instance), prepared from 1,1,3-trimethyl-diaminophenylindane and benzophenone-tetracarboxylic acid dianhydride, have been marketed over the last decade. Despite the partially aliphatic nature of polyimides containing the indane group, they show considerable retention of the thermal stability, with Tg values over 300 °C [107–110].

Apart from the indane group, heterocycles have been introduced in combination with imides in order to produce soluble aromatic copolyimides. While polyimides prepared from aromatic monomers containing heterocycles had high thermal stability but were insoluble in any organic solvent, [111, 112], some modern polyimides containing heterocyclic moieties like phenylquinoxaline [113], oxadiazole [114] or benzimidazole [115] show an acceptable degree of solubility. They all are soluble in organic polar solvents, show high Tg values, and excellent thermal resistance.

One of the most attractive and successful attempts in attaining processable aromatic polyimides is the introduction of fluorine atoms in the polymer structure, either as substituents of carbon atoms on the polymer backbone (as mentioned before for perfluoroalkane containing polyimides), or as perfluoromethyl or perfluoroalkyl side substituents. The most popular approach has been the introduction of the hexafluoroisopropylidene group in the main chain as a bulky separator group in the dianhydride monomer or in the diamine one [70].

The presence of trifluoromethyl groups and, in general, the substitution of fluorine for hydrogen, causes a dramatic change of properties. The combination of electronic and steric effects reduces the ability for interchain interactions and, particularly, hinders the formation of charge transfer complexes, which is a major factor of molecular packing and intractability in aromatic polyimides. Furthermore, the C-F bond is a high energy bond, so that polyimides containing fluorine are in general polymers with high Tg and excellent thermal properties, comparable to those of the conventional aromatic polyimides. They show some improved properties, such as

- low dielectric constant
- high optical transparency
- excellent mechanical properties
- low moisture absorption
- increased solubility
- low optical loss and low refractive index.

This excellent balance of properties has made fluorinated polyimides very attractive for some applications in advanced technologies, for instance

- high performance structural resins
- thermally stable coatings and films
- polymeric membranes for gases separation
- polymeric waveguides, and other electronic and optoelectronic applications.

Dozens of monomers, mainly diamines, have been described as suitable reactants for the synthesis of fluorinated polyimides in last years. Some examples have been listed in Table 6. Soluble and meltable polyimides have been prepared by combination of these with fluorinated and non-fluorinated monomers, and the range of structures achieved has already been so extended that it appears to be unlimited. They can be amorphous or semicrystalline, and have been processed by virtually any possible processing method.

Fluorinated polyimides have achieved great importance as barrier materials during the last few years. Many experimental polyimides prepared from fluorine-containing monomers, mainly novel diamines, show an advantageous balance of permeability and selectivity for technical gases and vapours, which makes them very attractive for the fabrication of permselective membranes [119]. This is an application field showing very rapid expansion, where there exists a strong demand for new polymeric materials, and where soluble aromatic polyimides are considered as a real alternative [136–146].

High-strength, high-modulus fibres are another modern application of aromatic polyimides. Early works on high temperature fibres had shown already that polyimide fibres can be manufactured by thermal cyclodehydration of wet spun polyamic acid fibres [13, 42, 147]. However, until the 1980s only one aromatic polyamide-imide fibre (Kermel) had undergone commercial development [148]. With the advent of fully imidized soluble polyimides, fibres of good mechanical properties could be fabricated from some particular thermoplastic polyimides [149, 150]. Nevertheless, the mechanical strength of such fibres is rather modest compared to other aromatic organic fibres, like aramides, that display strengths of about 3 GPa, and moduli of 120 GPa [68].

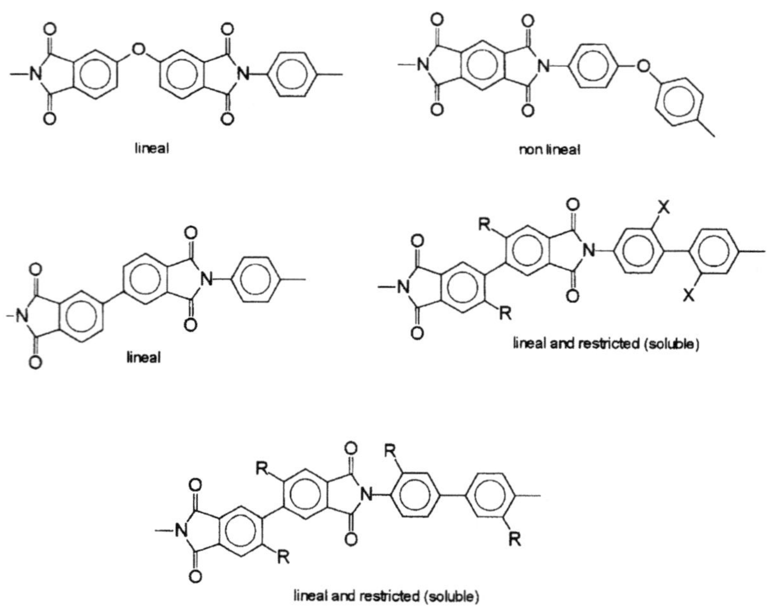

Table 6. Fluorinated monomers

Monomer	Reference
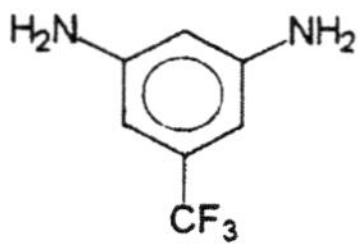	116, 117, 118
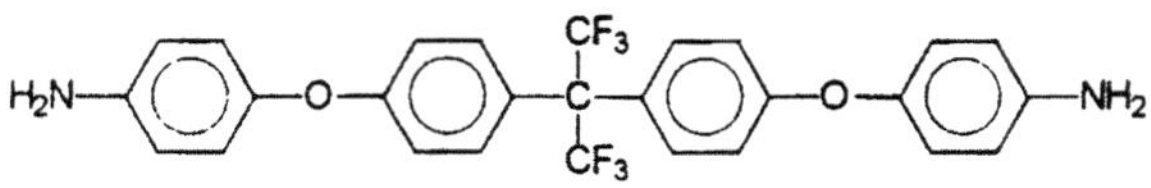	69, 76, 119, 120
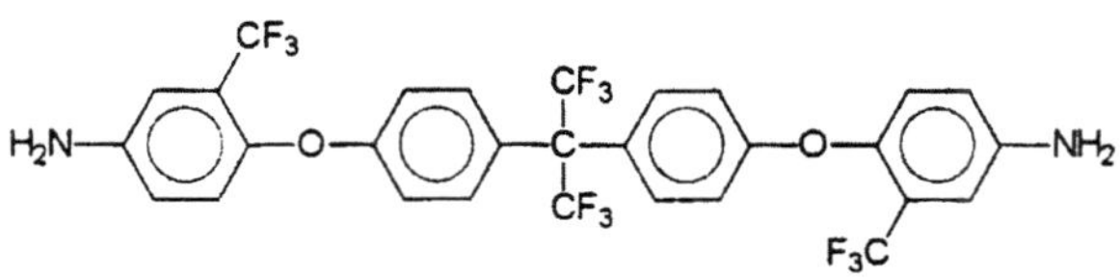	119
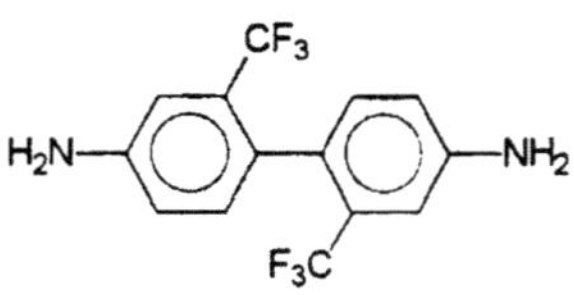	121, 122
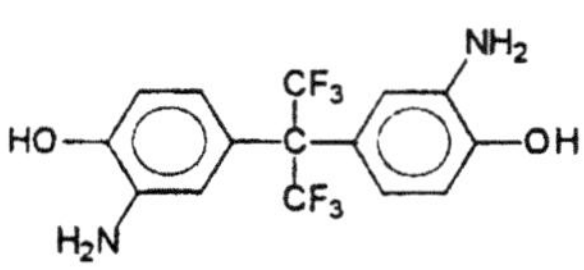	123
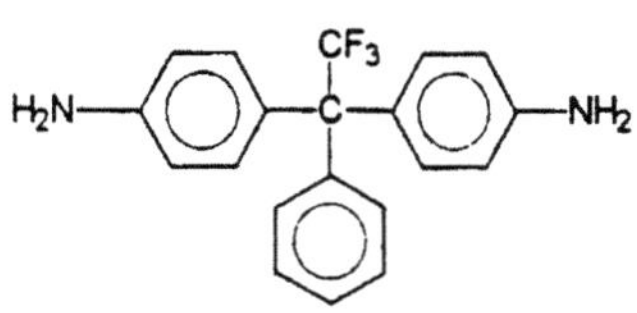	124, 125

Table 6. (continued)

Monomer	Reference
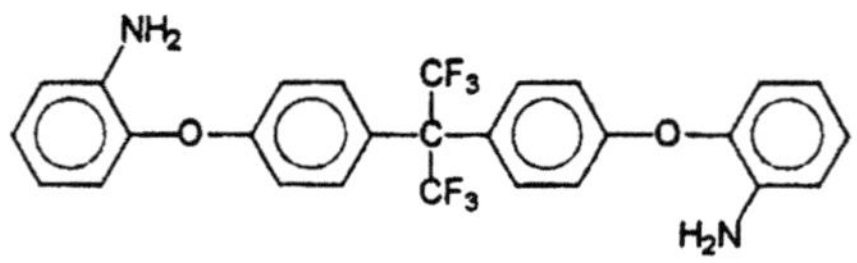	76
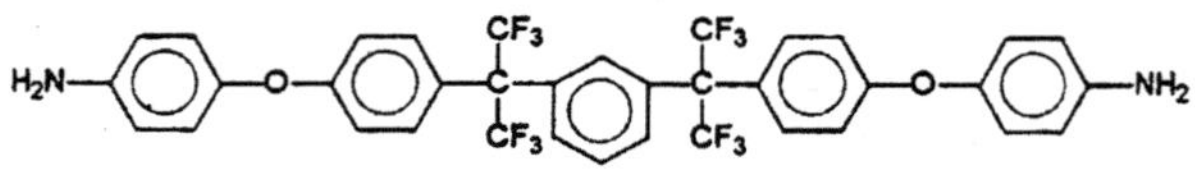	126
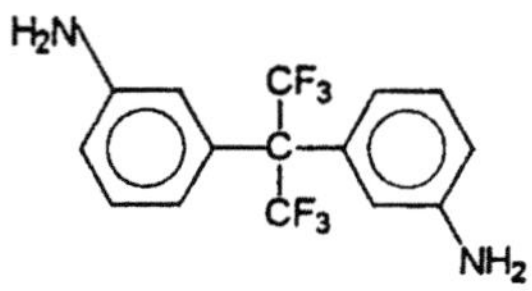	127, 128, 129
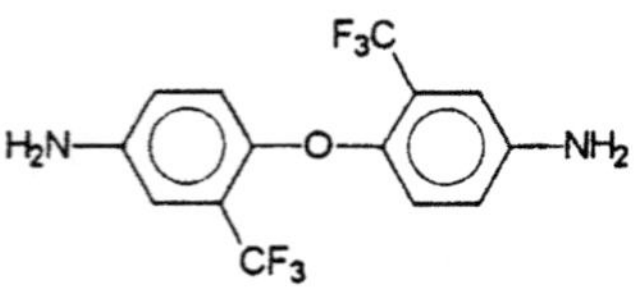	130
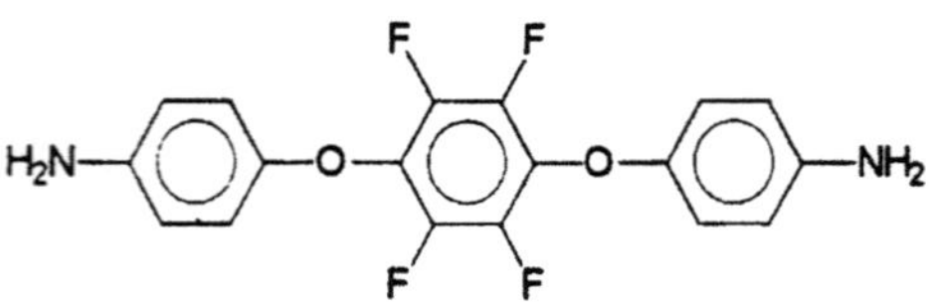	131
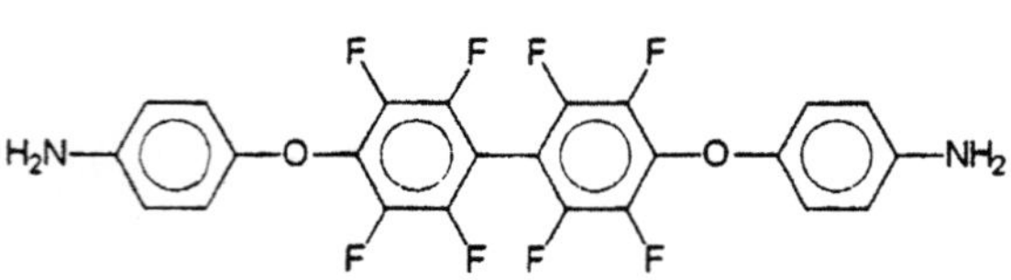	131

Table 6. (continued)

Monomer	Reference
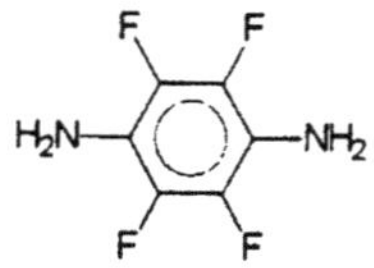	132
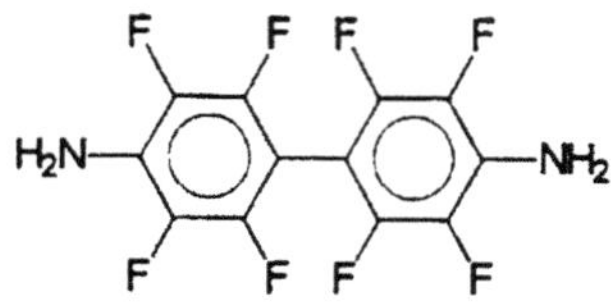	132
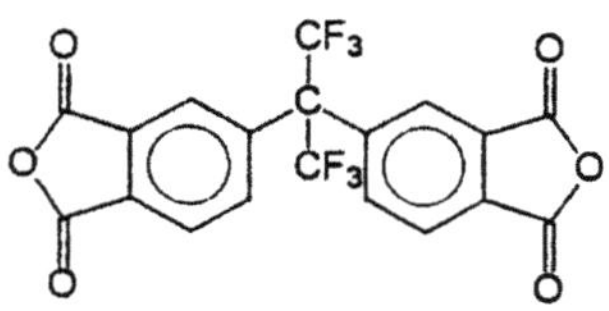	45, 119, 126, 133, 134
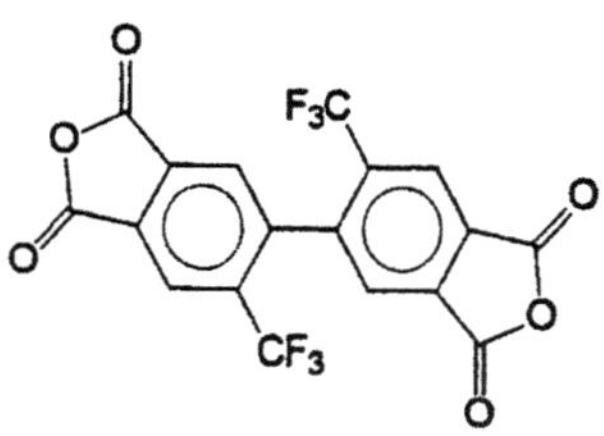	135

High-modulus, high-strength polyimide fibres have been prepared from monomers that can offer an extended chain conformation, along with enough solubility to be spun from solution, either by wet or by dry spinning [151, 152]. Thus, the combination of 3,3',4,4'-tetra-carboxydiphenyl dianhydride with *para*-oriented diamines (or diamines like 3,4'-diaminodiphenyl ether, that can adopt a lineal conformation) yields polyimides that can be wet spun in special conditions to fibres of up to 2 GPa strength and 100 GPa modulus [151, 152]. Polyimides based on 2,2'-disubstituted diphenyldianhydrides seem to be particularly suited for these purposes. Thus, fibres that display tensile strengths up to 3.2 GPa and moduli up to 130 GPa have been obtained from polyimides containing twisted biphenyl units along their backbones [135]. All these high-modulus, high-strength polyimide fibres were spun from solutions in *p*-chlorophenol, and then they were further drawn and annealed at high temperature (300–500 °C).

6 Commercial, Processable Aromatic Polyimides

In order to complete the range and to get a realistic picture about the actual importance of these materials, some emphasis should be placed on those polymers that have attained commercial or semi-commercial status.

Very few aromatic polyimides that are meltable can be processed in the molten state. For an aromatic polyimide to be processed in the melt, it must show an accessible melting temperature, with a decomposition temperature much higher than the melting temperature. Furthermore, the molten polymer should satisfy some requirements regarding viscosity and rheological characteristics, the molecular weight and the molecular weight distribution being crucial properties in this respect [44, 61, 153].

Not only do the chemical structure and the molecular weight affect the processability but also the method of synthesis, in particular the imidation step. Thermally imidized polyimides are always less tractable than solution imidized polyimides. That is because thermally imidized polyimides can undergo cross-linking, and because thermal treatment (about 300 °C) favour chains packing and provide higher molecular order than that achievable by solution imidation. Therefore, solution imidation is always preferable when thermoplastic polyimides are to be developed.

Molecular weight affects mainly the viscosity, so that, for some applications, such as adhesives that should be applied at relatively low temperatures, control of the molecular weight must be applied in the synthesis step [69, 154]. Controlled molecular weight polymers can be prepared by upsetting the stoichiometry

Table 7. Commercial processable aromatic polyimides

Name	Manufacturer	Remarks
ULTEM	General Electric	Bisphenol A based poly(ether-imide)
TORLON	Amoco	Poly(amide-imide)
AVIMID	DuPont de Nemours	Fluorine containing polyimide
LARC-TPI DURIMID	NASA Mitsui Toatsu	Poly(keto-imide)
New TPI AURUM	NASA Mitsui Toatsu	Poly(keto-ether-imide)
UPILEX	UBE	Biphenylene based polyimide
SIXEF	Höechst Celanese	Fluorine containing polyimide
MATRIMID	CIBA-Geigy	Indane containing polyimide
PI-84	Upjohn Lenzing. HPP	Poly(keto-imide) from diisocyanates
APICAL	Allied-Kanegafuchi	Biphenylene based polyimide
EYMYD	Ethyl Corporation TRW	Fluorine containing poly(ether-imide)

in favour of the dianhydride or the diamine monomer, or by adding a small amount of a monofunctional reactant, and these are the approaches frequently used to simplify the processing of thermoplastic polyimides [155–157].

With regards to crystallinity, aromatic polyimides are recognized as semirigid polymers that can crystallize in many instances, and, in fact, crystallinity has been reported for many polyimides. They can spontaneously crystallize when the chemical structure and the synthesis method are favourable, or they can develop crystallinity after an appropriate thermal treatment [65, 66, 74, 75].

Some semicrystalline aromatic polyimides have even achieved commercial development, and they can be processed by conventional means, with the precautions inherent to materials that do not flow below 300–380 °C. Nevertheless, these are exceptions, and the rule is to process in the melt essentially amorphous species, that flow under pressure over their Tg [65, 149, 158–161].

A list of commercial thermoplastic polyimides is given in Table 7, along with the name of the manufacturer and indicative information on the chemical composition.

7 Conclusion

The chemistry of aromatic polyimides has undergone outstanding development in last few years. Many novel species have been prepared by polycondensation of dianhydrides and diamines specially designed to overcome the traditional processing problems caused by the limited solubility and intractability of these high performance polymers. Efforts devoted to incorporating the structural elements that improve processability have led to an outstanding enrichment of the chemistry of polyimides, and have allowed the opening of new investigation and application areas for aromatic polyimides. Thus, fully imidized, aromatic polyimides can nowadays be molded from the melt like other high-temperature thermoplastics, they can be spun to high-strength, high-modulus fibres, can be cast as dense or microporous high efficiency, permselective membranes, or can be combined with reinforcing fibres for manufacturing high performance composites.

8 Appendix

As has been pointed out previously in this chapter, the processability of polyimides is strongly determined by the presence of single, rotatable linkages in the chain. These linkages have a double effect on the structure. On the one hand, these rotatable linkages increase the number of low-energy regions in conformational space, and therefore the presence of this kind of bond leads to an increase in conformational disorder and reduced persistence length and also difficults chain packing. On the other hand, the presence of flexible linkages causes a decrease of the barriers that separate minima. This decrease makes easier the conformational transitions, such as the glass transition and deformations.

Hence, a great number of monomers, either diamines or dianhydrides containing flexibilizing bridges between the rings, of the type -Ar-X-Ar- have been used for the synthesis of polyimides (see previous tables). The structure of X will determine the conformational behaviour of the bridges and therefore the properties of the polymers. Consequently, several attempts have been made to calculate, by computational simulation, the energy minima and the barriers between them as a function of the structure of X [163–167]. However, the strong dependence of the results obtained with the method used for the calculation means that the results obtained are in some cases contradictory.

Therefore, in this appendix, we have undertaken a comparative study on the conformational features of imide models of structure

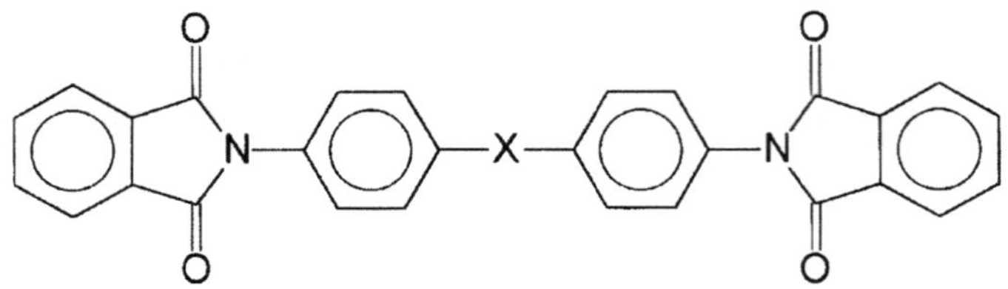

X being O; S, CH_2; $C(CH_3)_2$; $C(CF_3)_2$; CO and SO_2.

Molecular mechanics has been used to explore all the conformational possibilities of the different models. After this step, the precise position of the minima, and the value of the rotation barriers have been determined by quantum semiempirical methods, much more reliable than molecular mechanics calculations, mainly when there are strong electronic effects on the molecule.

The Dreiding 2.21 force field [168] has been used for the molecular mechanics calculations. Dreiding is a general force field well optimized for small molecules with C, H, N, O and S, that has been widely used for polymers [169, 170]. The total potential energy E was calculated as the sum of various energy contributions as follows.

$$E=E_{\text{bond stretch}}+E_{\text{bond bend}}+E_{\text{torsion}}+E_{\text{inversion}}+E_{\text{vdW}}+E_{\text{electrostatic}}$$

The AM1 method [171], included in the calculation package MOPAC-6 [172] has been used for the quantum semiempirical calculations.

The study has been performed by concerted analysis of the two rotation angles of both phenyl groups, ψ,Φ that were rotated simultaneously from 0 to 360° in 10° steps.

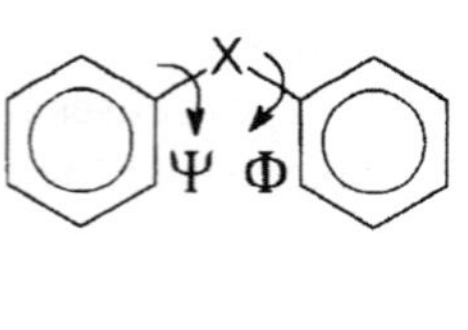

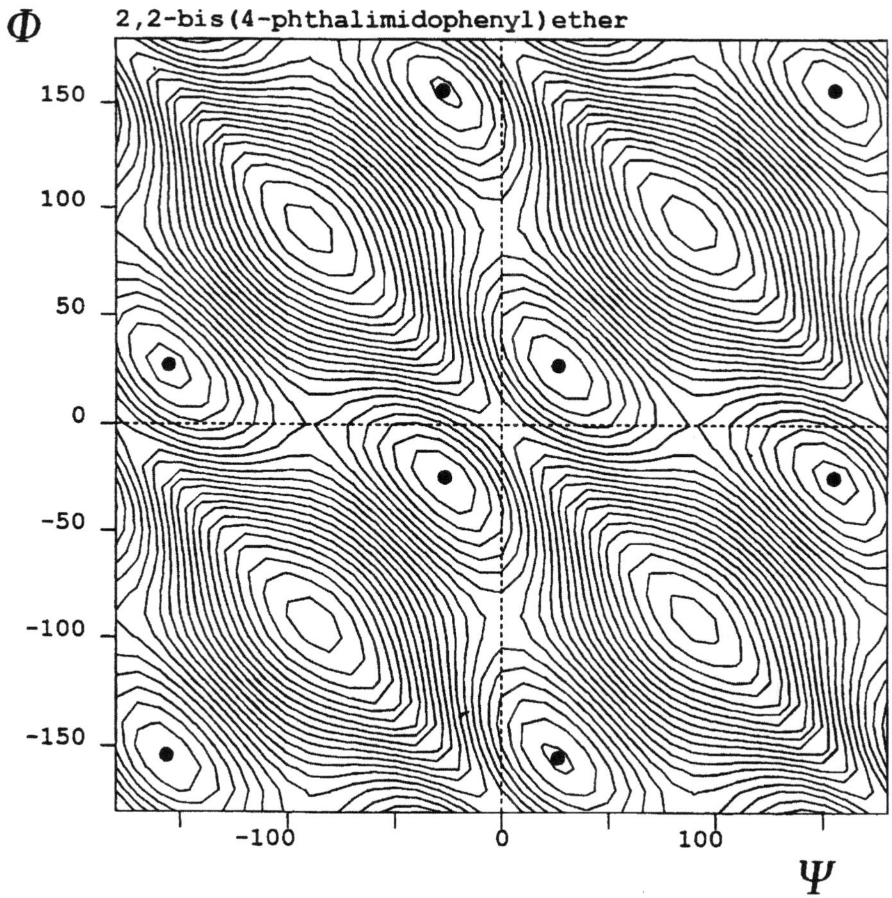

Fig. 2. Energy map for the 2,2-bis(4-phthalimidophenyl) ether. Contours are at 0.5 kcal/mol intervals above the minima (denoted by ●)

By this analysis, plots representing both angles and the corresponding energies can be obtained. As an example, Fig. 2 shows one of these plots for the case of 2,2-bis(4-phthalimidophenyl) ether. As can be seen, because of the symmetry of the aromatic rings, the plot has two planes of symmetry along the diagonals. Due to this symmetry, all the minima correspond to the same propeller conformation that has been observed several times in this kind of structure.

The saddle points between minima are located at 0°/±90° and ±90°/0°, although there is also another transition point at 0°/0°. Although this barrier seems to have the same energy as the others, when Dreiding is used, the use of semiempirical calculations shows that it is indeed a much higher energy barrier, due to the strong interaction between ortho hydrogens of both rings in the 0°/0° conformation. The use of another force field in this case (universal force field

Table 8. Position of the minima and energy of the rotational barriers of the different models, calculated by the AM1 semiempirical method

X	ψ/Φ at the minimum	$E_{transition}$-$E_{minimum}$ Kcal/mol
O	37°/37°	0.41
S	31°/31°	0.70
CH_2	69°/69°	0.32
$C(CH_3)_2$	53°/53°	0.60
$C(CF_3)_2$	48°/48°	1.08
CO	33°/33°	0.76
SO_2	90°/90°	2.33

[173]) gave results more similar to those obtained by AM1, but gave inconsistent results for the 2,2-bis(4-phthalimidophenyl)sulfone, indicating the strong dependence of the energy plots on the force field. That confirms the necessity of using quantum mechanical methods for this type of determinations. The energy barriers at 0°/±90° and ±90°/0° correspond to coordinated movements of both aromatic rings, so that the molecules move along low energy corridors connecting the different minima.

The position of the minima depends on the structure of X. For example, when X=SO_2, only four minima are found, located at ±90°/ and ±90°. The other models are placed between these extreme behaviours, with minima located at different angles. However the position of the transition points is always the same, at 0°/±90° and ±90°/0°, indicating that there is always a coordinated motion of the rings.

The quantum semiempirical modelling of the structures (AM1), without any restriction in the case of the minima, and restricting the rotation angle ψ to 0° in the transition structure, gives the data shown in Table 8. In this case, the coordinates of the minima were slightly different from those obtained by molecular mechanics while those of the transition points were always 0°/±90° and ±90°/0°, as predicted by molecular mechanics.

Very important differences were obtained, however, when comparing the energy of the rotational barriers obtained by both methods. Not only the absolute values were different, as was expected, because of the well known tendency of AM1 to underestimate the energy of the rotational barriers, but also the relative values of the energies. Thus, the energy of the barrier, determined by AM1, was higher when X=SO_2, contrary to that observed for molecular mechanics, probably due to difficulty of these methods to treat the electronic effects, that are very important in this case.

As said previously, the height of the barriers is directly related to the values of Tg. Therefore, a correlation can be tried between Tg and transition energy. To do that, the glass transition temperatures, taken from the literature, of a series of polyimides based on 6F-dianhydride and the diamines bearing the seven bridges we have calculated, have been represented vs the rotational barriers in Fig. 3. The great dispersity of Tg data found in the literature, that differ in some

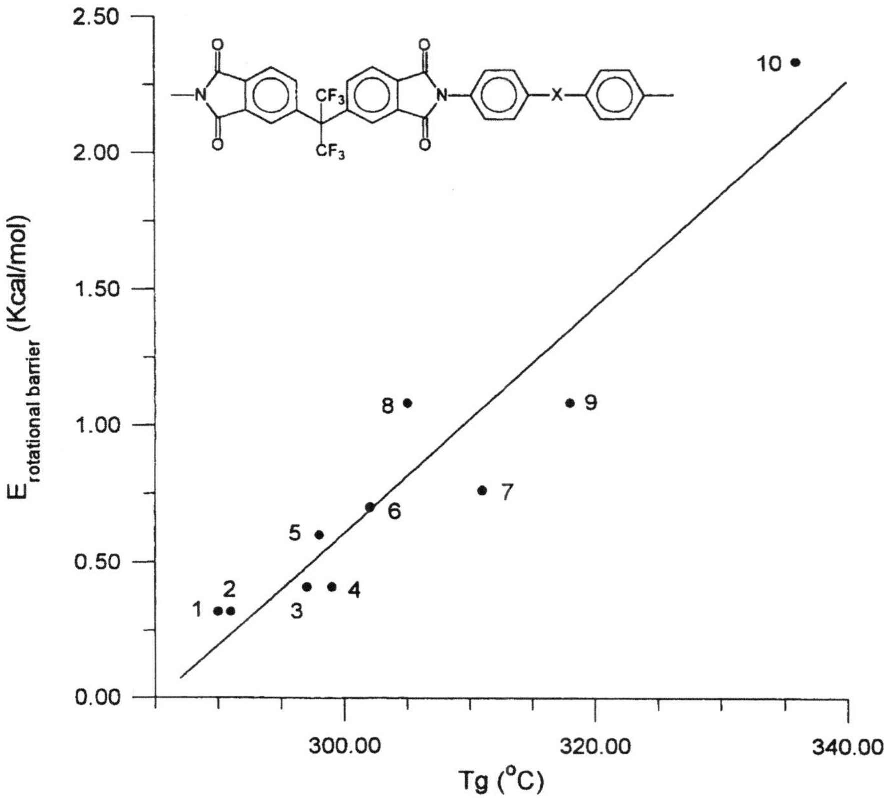

Fig. 3. Energy of the rotational barriers vs Tg of polyimides from 6F-dianhydride: 1, 2 – X= CH_2 [45, 70]; 3, 4 – X=O [141, 119]; 5 – X=$C(CH_3)_2$ [141]; 6 – X=S [162]; 7 – X=CO [45]; 8, 9 – X=$C(CF_3)_2$ [119, 129]; 10 – X=SO_2 [70]

cases by more than 40 °C, makes this type of representation difficult. In those cases, the higher and lower values were eliminated from the plot. A linear correlation can be found, indicating the existence of a good relationship between rotational barriers and Tg. This indicates that, in spite of the polar effects of X, the rotational barriers play a determining role in the value of Tg, and therefore on the processability of the polyimides.

9
References

1. Mittal KL (ed) (1985) Polyimides: synthesis, characterization and applications. Plenum, New York
2. Lee H, Stoffey D, Neville K (1967) New linear polymers. McGraw-Hill, New York
3. Sroog CE (1969) Polyimides. In: Mark F, Gaylord NG, Bikales NM (eds) Encyclopedia of polymer science and technology, vol 11. Interscience New York, p 247

4. de Abajo J (1992) Polyimides. In: Kricheldorf HR (ed) Handbook of polymer synthesis, vol 2. Marcel Dekker, New York, p 941
5. Alvino M, Frost LW (1971) J Polym Sci A-1 9:2209
6. Kinloch AJ (ed) (1986) Structural adhesives. Elsevier, London
7. May CA (ed) (1980) Resins for aerospace. ACS Symp Series No 132. Am Chem Soc, Washington
8. Serafini T, Delvigs P, Lithsey GR (1972) J Appl Polym Sci 16:905
9. Sat M (1997) Polyimides. In: Olabisi O (ed) Handbook of thermoplastics. Marcel Dekker, New York
10. Frazer AH (1968) High temperature resistant polymers. Interscience, New York
11. Cassidy PE (1980) Thermally stable polymers: synthesis and properties. Dekker, New York
12. Critchley JP, Knight GJ, Wright WW (1983) Heat-resistant polymers. Plenum, New York
13. Bessonov MT, Koton MM, Kudryavtsev VV, Laius LA, (1987) Polyimides: thermally stable polymers. Consultants Bureau, New York
14. Edwards WM, Robinson IM (1955) US Pat 2,710,853
15. Sroog CE (1967) J Polym Sci C 16:1191
16. Korschak VV, Babchinitser TM, Kazaryan LG, Vasilyev VA, Genin YV, Azriel AY, Vygodsky YS, Churochkina NA, Vinogradova SV, Tsvankin DY (1980) J Polym Sci Polym Phys 18:247
17. Bower GM, Frost LW (1963) J Polym Sci A 1:3135
18. Loncrini DF, Walton WL, Hugues RB (1966) J Polym Sci A-1 4:440
19. de Abajo J, Gabarda JP, Fontán J (1978) Angew Makromol Chem 71:143
20. Nieto JL, de la Campa JG, de Abajo J (1982) Makromol Chem 183:557
21. Yang C-P, Hsiao S-H (1989) Makromol Chem 190:2119
22. de Abajo J, de la Campa JG, Kricheldorf HR, Schwarz G (1990) Makromol Chem 191:537
23. Critchley JP, Grattan PA, White MA, Pippett JS (1972) J Polym Sci A-1 10:1789
24. de Visser AC, Gregonis DE, Driessen AA (1978) Makromol Chem 179:1855
25. Feld WA, Ramalingam B, Harris FW (1983) J Polym Sci Polym Chem 21:319
26. Padmanaban M, Toriumi M, Kakimoto M, Imai Y (1990) Makromol Chem Rapid Commun 11:15
27. Yilgör I, Yilgör Y, Eberle J, Steckle WP Jr, Johnson BC, Tyagi D, Wilkes GL, McGrath JE (1983), Polym Prep 24(1):170
28. Svetlichny VM, Denisov VM, Kudryatsev VV, Polotskaya GA, Kuznetsov YP (1991) In: Abadie MJM, Sillion B (eds) Polyimides and other high-temperature polymers. Elsevier, Amsterdam, p 525
29. Buba GN (1985), In: Mittal KL (ed) Polyimides: synthesis, characterization and applications. Plenum, New York, p 51
30. Critchley JP, White MA (1972) J Polym Sci A-1 10:1809
31. Kricheldorf HR, Linzer V (1995) Polymer 36:1893
32. Marek M Jr, Doskocilová D, Schmidt P, Schneider B, Kriz J, Labsk_ J, Puffr R (1994) Polymer 35:4881
33. Cheng SZD, Mittleman ML, Janimak JJ, Shen D, Calmers TM, Lien, H, Tso CC, Gabori, PA, Harris, FW (1992) Polym Prep 33(1):449
34. Acevedo M, Harris FW (1994) Polymer 35:4456
35. Koning CE, Teuwen L, Meijer EW, Moonen Y (1994) Polymer 35:488
36. Labadie JW, Hedrick JL (1990) SAMPE J, 26:19
37. Padmanaban M, Toriumi M, Kakimoto M-A, Imai Y (1990) J Polym Sci A Polym Chem 28:3261
38. Rogers ME, Glass TE, Mechan SJ, Rodriguez D, Wilkes GL, McGrath JE (1994) J Polym Sci A Polym Chem 32:2663
39. Moon YD, Lee YM (1993) J Appl Polym Sci 50:1461
40. John T, Valenty VB (1988) Proccedings 3rd Int Conference on Polyimides, Ellenville, p 36
41. Stern SA, Liu Y, Feld A (1993) J Polym Sci B Polym Phys 31:939

42. Adrova N, Bessonov M, Laius LA, Rudakov AP (1969) Polyimides: a new class of heat-resistant polymers. IPST Press, Jerusalem
43. Dine-Hart RA, Wright WW (1972) Makromol Chem 153:237
44. Gibbs HH, Breder CV (1974) Polym Prep 15(1):775
45. St. Clair TL, St. Clair AK, Smith EN (1976) Polym Prep 17(2):359
46. Takekoshi T, Wirth JG, Heath DR, Kochanowski JE, Manello JS, Webber MJ (1980) J Polym Sci Polym Chem 18:3069
47. Davies M, Hay JN, Woodfine B (1992) High Perform Polym 5:37
48. St Clair TL, St Clair AK, Smith EN (1977) In: Harris FW, Seymour RB (eds) Structure-solubility relationships in polymers. Academic Press, New York, p 199
49. Huang W, Tong YT, Xu J, Ding M (1997) J Polym Sci, A Polym Chem 35:143
50. Bell VL, Stump BL, Gager H (1976) J Polym Sci Polym Chem 14:2275
51. Lin Y-N, Joardar S, McGrath JE (1993) Polym Prep 34(1):515
52. Malinge J, Garapon J, Sillion B (1988) Brit Polym J 20:431
53. Lin T, Stickney KW, Rogers M, Riffle JS, McGrath JE, Yu TH, Davis RM (1993) Polymer 34:772
54. Gerber MK, Pratt JR, St Clair TL (1988) Proceedings 3rd Int Symp on Polyimides, Ellenville, p 101
55. Connel JW, Croall CI, Hergenrother PW (1992) Polym Prep 33(1):1101
56. Kawashima Y, Ikeda T, Kitagawa H (1988) Proceedings 3rd Int Symp on Polyimides, Ellenville, p 32
57. Strong DB (1993) High performance and engineering thermoplastic composites. Technomic Pub, Lancaster PA
58. Weinrotter K, Vodiunig R (1989). In: Feger C, Khojasteh MM, McGrath JE (eds) Polyimides: materials, chemistry and characterization. Elsevier, Amsterdam
59. Takekoshi T (1990) Adv Polym Sci 94:1
60. Abadie MJM, Sillion B (1991) Polyimides and other high-temperature polymers. Elsevier, Amsterdam
61. Yamaguchi A, Ohta M (1993) SAMPE J 23:28
62. Moy TM, DePorter CD, McGrath JE (1993) Polymer 34:819
63. Burks HD, St Clair TL (1985) J Appl Polym Sci 30:2401
64. Eastmond GC, Prapotny J (1994) Polymer 35:5148
65. Takahashi T, Yuasa S, Tsuji M, Sakurai K (1994) J Macromol Sci Phys B33(1):63
66. Huo PP, Friler JB, Cebe P (1993) Polymer 34:4387
67. Brode GL, Kwiatkowski GT, Kawakami J (1974) Polym Prepr 15(1):761
68. Yang HH (1989) Aromatic high-strength fibers. Interscience, New York
69. Scola DA, Pike RA, Vontell JH, Pinto JP, Brunette CM (1989) High Perform Polym 1:17
70. Cassidy PE, Aminabhavi TM, Farley JM (1989) J Macromol Sci Rev C29:365
71. Young PR, Davis JRJ, Chang AH, Richardson JN (1990) J Polym Sci A Polym Chem 28:3107
72. Zoia G, Stern SA, St Clair AK, Pratt JA (1994) J Polym Sci B Polym Phys 32:53
73. Bell VL, Kilzer L, Hett EM, Stokes GM (1981) J Appl Polym Sci 26:3805
74. Havens SJ, Hergenrother PM (1990) J Polym Sci A Polym Chem 28:2427
75. Muellerleile JT, Risch BG, Rodriguez DE, Wilkes GL (1993) Polymer 34:789
76. Asanuma TA, Oikawa H, Ookawa Y, Yamasita W, Matsuo M, Yamaguchi A (1994) J Polym Sci A Polym Chem 32:2111
77. Liou G-S, Maruyama M, Kakimoto M-A, Imai Y (1993) J Polym Sci A Polym Chem 31:3273
78. Kumar D, Gupta AD (1995) Polym Prep 36(2):49
79. Yang C-P, Lin J-H (1995) J Polym Sci A Polym Chem 31:2153
80. Davies M, Hay JN, Woodfine B (1993) High Perform Polym 5:37
81. Pratt JR, Blackwell DA, St Clair TL, Allphin NL (1989) Polym Eng Sci 29:63
82. Eastmond GC, Paprotny J, Webster I (1993) Polymer 34:2865
83. Sek D, Pijet P, Wanic A (1992) Polymer 33:190

84. Rusanov AL (1994) Adv Polym Sci 111:177
85. Korshak VV, Rusanov AL (1969) Izv Akad Nauk SSSr Ser Khim 10:2418; through CA 70, 2054s
86. Rusanov AL, Komarova LG, Sheveleva TS, Prigozhina MP, Shevelev SA, Dutov MD, Vatsadze LA, Serushkina OV (1996) Proc Int Symp on Polycondensation, Paris, p 75
87. Rusanov AL, Shifrina ZB (1993) High Perform Polym 5:107
88. Harris FW, Norris SO (1973) J Polym Sci Polym Chem Ed 11:2143
89. Harris FW, Hsu SL-C (1989) High Perform Polym 1:3
90. Giesa R, Keller U, Eiselt P, Schmidt H-W (1993) J Polym Sci A Polym Chem 31:141
91. García C (1996) Thesis, Autonoma Univ, Madrid
92. Imai Y, Maldar NN, Kakimoto M-A (1984) J Polym Sci Polym Chem 22:2189
93. Akutsu F, Kataoka T, Shimizu H, Naruchi K, Miura M (1994) Macromol Chem Rapid Commun 15:411
94. Spiliopoulos IK, Mikroyannidis JA (1996) Macromolecules 29:5313
95. Korshak VV, Vinogradova SV, Vygodskii YS (1974) J Macromol Sci Rev Macromol Chem C11:45
96. Korshak VV, Vinogradova SV, VygodskiiYS, Nagiev ZM, Urman YG, Alekseeva SG, Slonium IY (1983) Makromol Chem 184:235
97. Lozano AE, de la Campa JG, de Abajo J, Preston J (1993) J Polym Sci A Polym Chem 31:138
98. Lozano AE, de la Campa JG, de Abajo J, Preston J (1994) Polymer 35:873
99. Kricheldorf HR, Schwarz G (1989) Makromol Chem Rapid Commun 10:243
100. Tsai T, Arnold FE (1986) Polym Prep 27(2):221
101. Ayala D, Lozano AE, de la Campa JG, de Abajo J (1997) Polym Prep 38(2):359
102. Jeong H-J, Oishi Y, Kakimoto M-A, Imai Y (1991) J Polym Sci A Polym Chem 29:39
103. Becker KH, Schmidt H-W (1992) Macromolecules 25:6784
104. Harris FW, Sakaguchi Y (1988) Proc 3rd Int Conference on Polyimides, Ellenville, p 25
105. Oishi Y, Ishida M, Kakimoto M-A, Imai Y (1992) J Polym Sci A Polym Chem 30:1027
106. Auman BC, Higley DP, Scherer KV Jr, McCord EF, Shaw WH Jr (1995) Polymer 36:651
107. Falcigno P, Masola M, Williams D, Jasne S (1988) Proc 3rd Int Conference on Polyimides, Ellenville, p 83
108. Wusik MJ, Jha B, King M (1991) Polym Prep 32(3):218
109. Chao HS-I, Barren E (1992) Polym Prep 33(1):1024
110. Maier D, Yang D, Wolf M, Nuyken O (1994) High Perform Polym 6:335
111. Preston J, Dewinter WF, Black WB (1969) J Polym Sci A-1 7:283
112. Preston J, Dewinter WF, Black WB, Hofferbert WL Jr (1969) J Polym Sci A-1 7:3027
113. Sava I, Bruma M, Belomoina N, Mercer FW (1993) Angew Makromol Chem 211:113
114. Mercer FW, Mckenzie MT (1993) High Perform Polym 5:97
115. Ponomarev II, Nikol'skii OG, Volkova YA, Zakharov AV (1994) Polym Sci Ser A 36:1185
116. Gerber MK, Pratt JR, St Clair AK, St Clair TL (1990) Polym Prep 31(1):340
117. Buchanan RA, Mundhenke RF, Lin HC (1991) Polym Prep 32(2):193
118. Jensen BJ, Havens SJ (1992) Polym Prep 33(1):1084
119. Tanaka K, Kita H, Okano M, Okamoto K-I (1992) Polymer, 33:585
120. Jones RJ, Silverman EM (1989) SAMPE J 25:41
121. Ando S, Matsuura T, Nishi S (1992) Polymer 33:2934
122. Matsuura T, Ando S, Saki S, Yamamoto F (1994) Macromolecules 27:6665
123. Omote T, Koseki K, Yamaoka T (1990) J Polym Sci C Polym Lett 28:59
124. McGrath JE, Grubbs H, Rogers ME, Mercier R, Joseph WA, Alston W, Rodríguez D, Wilkes GL (1992) Polym Prep 33(1):445
125. Rogers ME, Woodard MH, Brennan A, Cham PM, Marand H, McGrath JE (1992) Polym Prep 33(1):461
126. Cassidy PE, Aminabhavi TM, Farley JM (1989) J Macromol Sci Macromol Rev C29:365
127. Husk RS, Cassidy PE, Gebert KL (1988) Macromolecules 21:1234
128. Feger C (1991) Polym Prep 32(2):76
129. Chung T-S, Vora RH, Jaffe M (1991) J Polym Sci A Polym Chem 29:1207

130. Buchanan RA, Mundhenke RF, Lin HC (1991) Polym Prep 32(2):193
131. Misra AC, Tesoro G, Hougham G, Pendharkar S (1992) Polymer 33:1078
132. Hougham G, Shaw J, Tesoro G (1988) Proc 3rd Int Conference on Polyimides, Ellenville, p 40
133. Scola DA (1991) In: Abadie MJM, Sillion B (eds) Polyimides and other high-temperature polymers. Elsevier, Amsterdam, p 265
134. Matsuura T, Hasuda Y, Nishi S, Yamada N (1991) Macromolecules 24:5001
135. Harris FW, Li F, Lin S-H, Chen J-C, Cheng SZD (1997) Macromol Symp 122:33
136. Koros WJ, Fleming GK, Jordan SM, Kim TH, Hohen H (1988) Prog Polym Sci 13:339
137. Ghosal K, Freeman BD (1994) Polymers Adv Tech 5:673
138. Kim TH, Koros WJ, Husk GR, O'Brien KC (1987) J Appl Polym Sci 34:1767
139. Yamamoto H, Mi Y, Stern SA, St Clair AK (1990) J Polym Sci B Polym Phys 28:2291
140. Eastmond GC, Page PCB, Paprotny J, Richards RE, Shaunak R (1993) Polymer 34:667
141. Matsumoto K, Xu P (1993) J Appl Polym Sci 47:1961
142. Langsam M, Burgoyne WF (1993) J Polym Sci A Polym Chem 31:909
143. Stern SA (1994) J Membrane Sci 94:1
144. Fritsch D, Peinemann KV (1995) J Membrane Sci 99:29
145. Kawakami H, Anzai JS (1995) J Appl Polym Sci 57:789
146. Kawakami H, Mikawa M, Nagaoka S (1997) J Membrane Sci 137:241
147. Irwin RSW (1967) J Polym Sci C 19:77
148. Pigeon R, Allard P (1974) Ang Makromol Chem 40/41:139
149. Serfaty W (1985) Polyetherimide. In: Margolis JM (ed) Engineering thermoplastics. Dekker, New York
150. Weinrother K (1988) Proc Nonwoven Conference, Atlanta
151. Irwin RS (1984) Polym Prep 25(2):213
152. Kaneda T, Katsura T, Nakagawa K, Makino H (1986) J Appl Polym Sci 32:3151
153. Yamaguchi A, Ohta M (1987) SAMPE J 23:28
154. Burks HD, St Clair TL, Gautreaux CR (1990) SAMPE J 26:59
155. Progar D, St Clair TL, Gautreaux C, Yamaguchi A, Ohta M (1989) Int SAMPE Symp 21:544
156. Pratt JR, St Clair TL (1990) SAMPE J 26:29
157. Hergenrother PM, Havens SJ (1993) High Perform Polym 5:177
158. Schobesberger M (1992) Int Diamanten Rundschau 26:2
159. Schobesberger M (1994) Kunststoffe 80:759
160. Bystry-King FA, King JJ (1985) Polyimides. In: Margolis JM (ed) Engineering thermoplastics. Dekker, New York
161. Billerbeck CJ, Henke SJ (1985) Torlon poly(amide imide). In: Margolis JM (ed) Engineering thermoplastics. Dekker, New York
162. Wessling M (1993) Ph. D. Thesis, University of Twente
163. Anwer A, Lowell R, Windle AH (1989) Polym Prepr 30(2):86
164. Kendrick JJ (1990) Chem Soc Faraday Trans 86(24):3995
165. Fan CF, Hsu SL (1991) Macromolecules 24:6244
166. Chen CL, Chen HL, Lee CL, Shih JH (1994) Macromolecules 27:2087
167. Fan CF, Cagin T, Chen ZM, Smith KA (1994) Macromolecules 27:2383
168. Mayo SL, Olafson BD, Godard WA III (1990) J Phys Chem 94:8897
169. Zhang R, Mattice WL (1995) Macromolecules 28:7454
170. Vasudevan VJ, McGrath JE (1996) Macromolecules 29:637
171. Dewar MJS, Zoebisch EG, Healy EF, Stewart JJP (1985) J Am Chem Soc 107:3902
172. MOPAC 6.0 (1990) Quant Chem Prog Exch 455
173. Rappé AK, Casewit CJ, Colwell KS, Goddard WA, Skiff WM (1992) J Am Chem Soc 114:10,024

Received: March 1998

Synthesis and Characterization of Segmented Polyimide-Polyorganosiloxane Copolymers

James E. McGrath, Debra L. Dunson, Sue J. Mecham[1], and James L. Hedrick[2]

[1] Department of Chemistry and Center for High Performance Polymeric Adhesives and Composites, 2108 Hahn Hall, Virginia Tech, Blacksburg, VA 24061-0344, USA
[2] IBM Almaden Research Center, 650 Harry Road, San Jose, CA 95120-6099, USA

Polyimide-polydimethylsiloxane block or segmented copolymers are reviewed with respect to synthesis, characterization, structure-property relationships and other special characteristics. The siloxane-modified imides have developed into important materials over the past 10 years. Although some thermal and thermooxidative stability is lost through introduction of the siloxane segment, a number of improvements are also observed, in processibility, toughness, flexibility, adhesion and membrane performance, to mention a few. The chemistry involves first preparing an organo-functional siloxane oligomer of a few thousand molecular weight, by ring-opening equilibration processes in the presence of an end blocker to produce either aminopropyl or anhydride end-groups. The resulting oligomers are then usually reacted with the two imide-forming monomers to generate "randomly segmented copolymers". If the siloxane block lengths are designed to exceed even 1000 $\overline{Mn}$, a microphase separation allows for the generation of two glass-transition temperatures and other desirable multiphase bulk properties, which are compositionally dependent. In addition, the hydrophobic nature of the siloxane microphase influences the surface properties, often in desirable ways, useful for aerospace and microelectronics, and these features are discussed in this review.

Keywords. Polyimides, Polydimethylsiloxane, Block or segmented copolymers, Synthesis Characterization, Physical Behavior, Adhesion, Membranes, Review

Advances in Polymer Science, Vol.140

1 Introduction and Scope

1.1 Importance of This Subject

Polyimides are arguably the most thermally stable organic polymeric materials that have been discovered and are of great interest for many high performance applications [1–8]. However, these polymers may be insoluble and intractable in their fully imidized form unless they are carefully designed. Therefore, much effort has been spent developing the synthesis of processable polyimides that maintain reasonably high strength and thermooxidative stability. Several successful strategies have now been made and are discussed elsewhere within this monograph. This chapter focuses on the incorporation of flexible polydimethylsiloxane oligomeric segments into a polyimide backbone, which often yields soluble processable segmented polymers with a variety of good property characteristics. The polydimethylsiloxanes, in particular, are known to impart a number of beneficial properties, which include enhanced solubility, UV stability, resistance to degradation in aggressive oxygen environments, impact resistance, modified surface properties and a number of other interesting features such as good adhesion, low stress in thin film dielectrics, semi-conducting polymer electrolytes, fire resistance, gas separation membranes and thermoplastic elastomeric applications. The subject has thus far developed into an important field with many potential applications, and this chapter provides an expansion and update of our earlier review [9], which is now already 10 years old.

1.2 Historical Perspectives

Block copolymers were developed rapidly in the 1960s when living anionic polymerization was first utilized to synthesize triblock thermoplastic elastomers or elastoplastics. At the same time, step or condensation polymerization to produce thermoplastic polyurethanes, urea-urethane spandex fibers, and later more specialized materials, such as the semicrystalline polyester-polyether copolymers were developed [10]. Imide block or segmented copolymers utilizing polydimethylsiloxane systems were somewhat slower to develop. Some of the first publications came from the laboratory of Greber [11,12], Kucklertz [13] and in the patent literature from Holub and his colleagues [14,15] at General Electric. Although oligomeric siloxanes had already been introduced into multiblock or segmented polymer structures such as the polycarbonate or polyether sulfones [10], translation into the imide systems was not immediate. Efforts at NASA Langley by St. Clair and his colleagues in the early 1980s [16,17] and from our laboratory [18–22] were some of the first contributions reflecting the desire to establish multiphase block copolymers comprised of polydimethylsiloxane and polyimides. Indeed, in the review by Yilgor and McGrath, which appeared in 1988 [9], hardly any publications had yet appeared in the open literature that included reactive oligomeric siloxanes into segmented polyimide copolymers, even though thousands of papers were available on siloxane chemistry and even a large number of other polydimethylsiloxane block copolymers. Thus, this current review will focus on the synthesis and characterization of segmented polyimide-polydimethylsiloxane copolymers and will review the various methods that have been used to prepare the appropriate silicon-carbon bonded reactive organofunctional siloxanes.

Principally, aminopropyl and aryl or cycloaliphatic anhydride functional systems, along with some effort on aminophenyl functional oligomers have been employed. The multiphase segmented copolymers are normally prepared by reacting the polydimethylsiloxane oligomers under homogeneous conditions with the two imide-forming comonomers, i.e. a dianhydride and a diamine. Alternate chemistries involving, for example, transimidization or amic ester coupling strategies are now also well known. Many of the early imide-siloxane materials were based on relatively rigid hard blocks derived from, for example, pyromellitic dianhydride (PMDA) or biphenyl dianhydride (BPDA) and para- or meta-phenylenediamine (PDA) or oxydianiline (ODA).

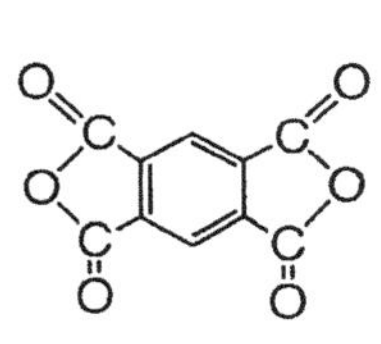

PMDA

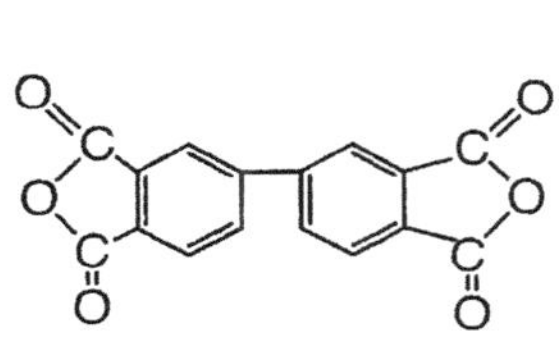

BPDA

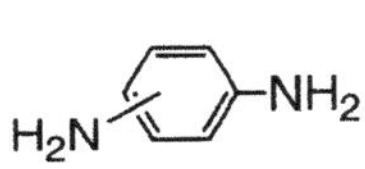

PDA

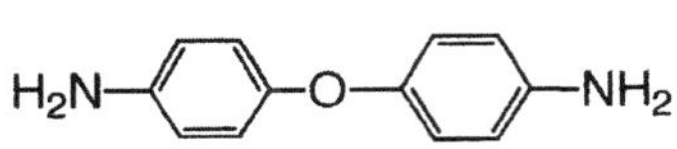

ODA

In general, these materials produced insoluble systems after imidization and had to be processed by first casting the amic acid intermediate onto a substrate, then imidizing. More recently [8], many examples have been demonstrated where the resulting polyimide is thermoplastic, melt processable and thus can

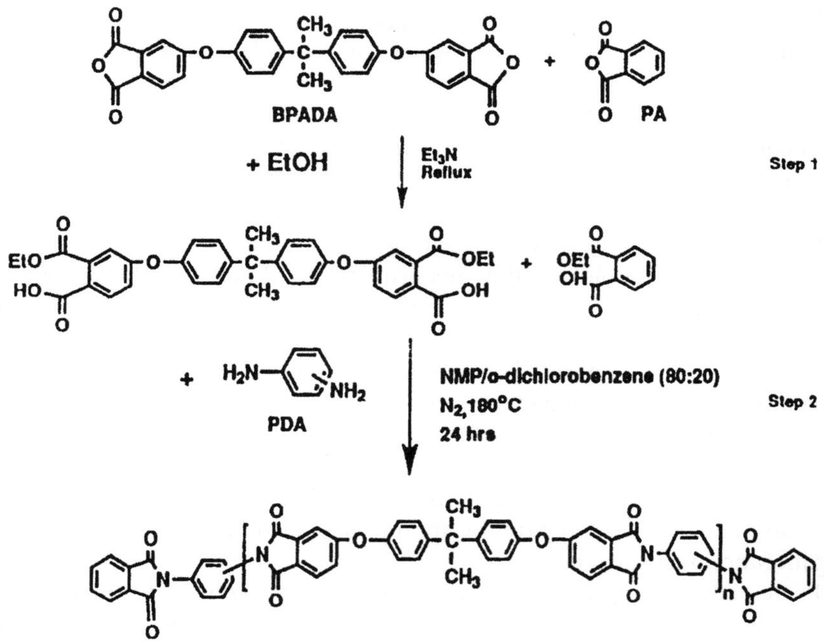

Scheme 1. Synthesis of BPADA/PDA PA capped thermoplastic polyetherimide by the acid-ester route

either be prepared from polyamic acid solution casting, or by preimidizing and subsequently fabricating from the melt. Considerable effort in the General Electric laboratories based on the Ultem type polyether imide has been reported, since it is known to be a soluble thermoplastic processable material [23–30]. An illustrative ester-acid route to this structure is provided in Scheme 1. These approaches will be reviewed in this chapter along with strategies that have been used to produce perfectly alternating block copolymers utilizing techniques such as transimidization. Structure-property relationships are of great current interest and many examples have been found during the literature search that show the applicability of these relatively new copolymers in a variety of material science areas, as mentioned earlier.

2 Synthesis

2.1 Organofunctional Oligomers

The synthetic chemistry for organofunctional oligomers generally follows one of two routes. Either the materials can be prepared by ring-opening polymeriza-

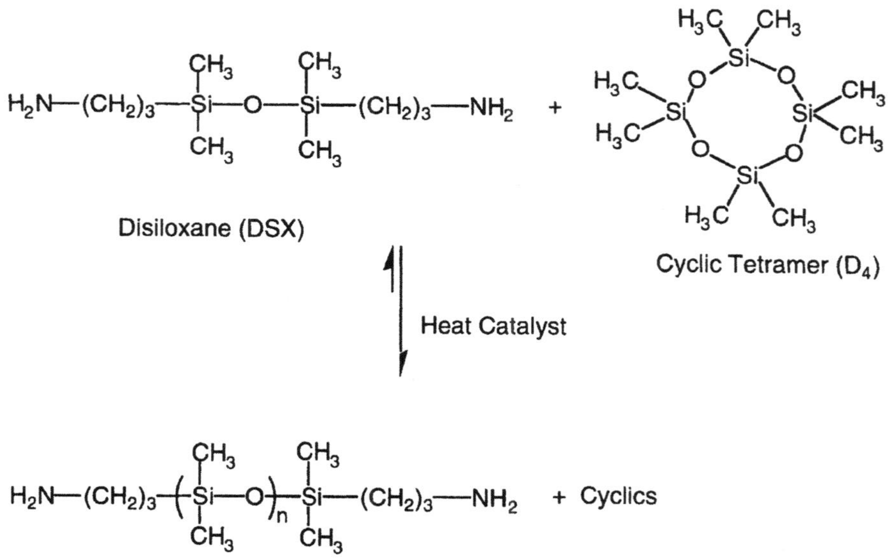

Scheme 2. Synthesis of polysiloxane oligomers [36]

tion of most commonly the cyclic tetramer (D_4) in the presence of siloxane chain transfer agents containing silicon–carbon bonds, called end-blockers, or the resulting materials may be generated by end-capping oligomeric "silicone" (polydimethylsiloxane) fluids with cyclic precursors of the end-capper. The structures most commonly used have employed aminopropyl functional oligomers, such as those shown in Scheme 2.

The aminopropyl functional polydimethylsiloxane oligomers have been successfully synthesized in our laboratory and elsewhere by the ring-opening polymerization of the cyclic tetramer D_4 in the presence of the disiloxane (DSX). The resulting dimer was prepared by the hydrosilation of dimethylchlorosilane with protected allylamine, followed by deprotection, hydrolysis and coupling, to yield the disiloxane, as illustrated in Scheme 3 [31]. A variety of basic catalysts have been used and although potassium hydroxide may have some commercial advantages, tetramethylammonium hydroxide has been successfully utilized by a variety of investigators [32]. The quaternary hydroxide generates the analogous siloxanolate, which then effectively attacks the silicon ring to open the cyclic tetramer at 80 °C. The disiloxane serves as a chain end-blocker since the silicon–carbon bonds are covalent, but the silicon–oxygen bonds are partially ionic, and thus can participate in the reaction and effectively produce the final difunctional oligomer. The polymerization of the tetramer is well known to have almost no reaction enthalpy [33,34] and, hence, a ring chain equilibrium is formed, which produces about 85% of the desired linear oligomer. The remaining cyclics can be easily vacuum stripped and the quaternary catalyst can be de-

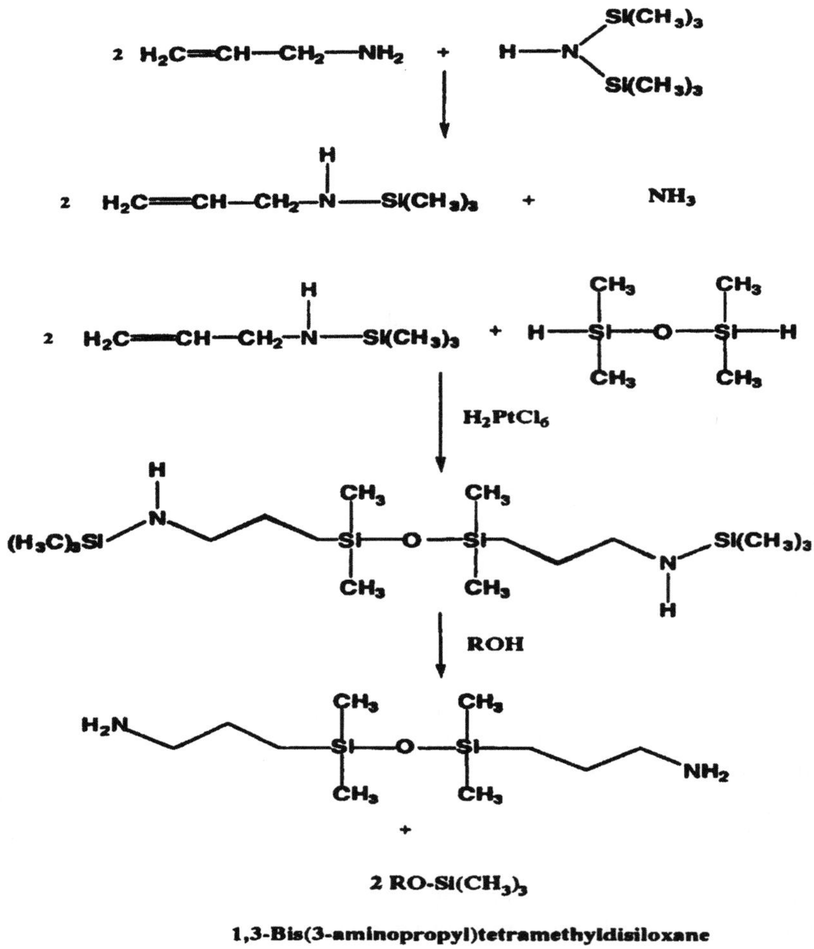

Scheme 3. Synthesis of 1,3-bis(3-aminopropyl)tetramethyldisiloxane [31]

composed by a brief treatment at about 150 °C to produce a neutralized, stable oligomer [35–37]. The number-average molecular weight at equilibrium, Mn, is controlled by the initial ratio of the cyclic tetramer to the disiloxane. One can titrate the aminopropyl end-groups or analyze the oligomers by proton, carbon or ^{29}Si NMR spectra to assess functionality and molecular weight. The utilization of these techniques is illustrated below and assignments are provided for both the proton NMR (Fig. 1), ^{13}C NMR (Fig. 2), and ^{29}Si NMR (Fig. 3). The agreement between the three spectroscopic methods is excellent, not only for laboratory samples, but for even commercially available materials. Infrared analysis (Fig. 4) is also useful for characterization.

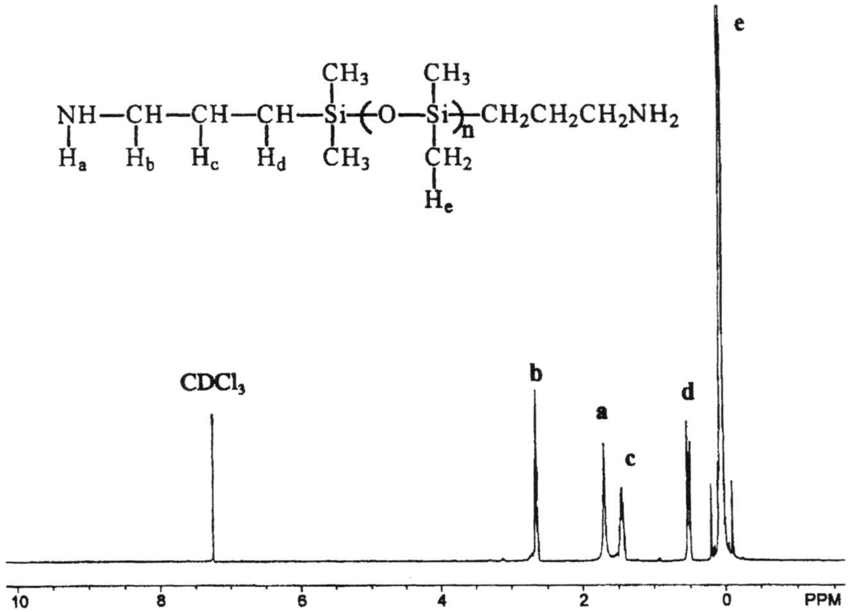

Fig. 1. 400 MHz ^1H NMR spectrum of primary aminoalkyl-terminated polydimethylsiloxane (PDMS) (Mn=1528)

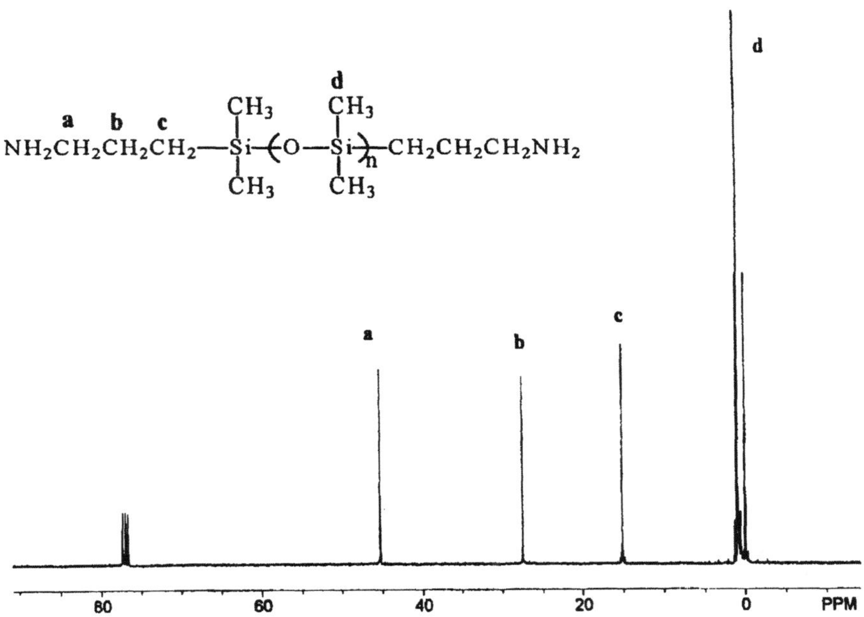

Fig. 2. 100 MHz ^{13}C NMR spectrum of primary aminoalkyl-terminated PDMS (Mn=1528)

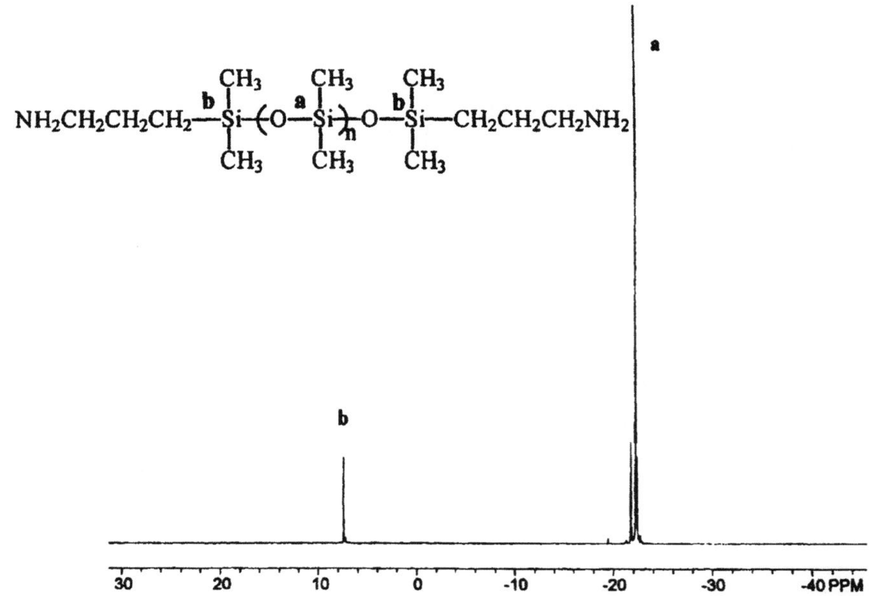

Fig. 3. 79 MHz ^{29}Si NMR spectrum of primary aminoalkyl-terminated PDMS (Mn=1528)

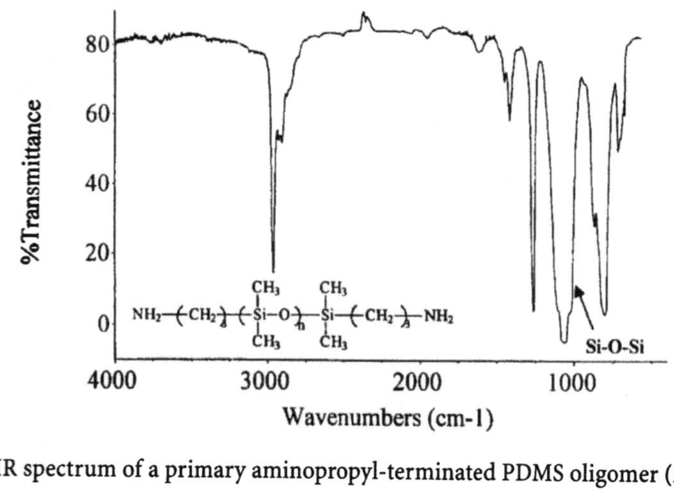

Fig. 4. FTIR spectrum of a primary aminopropyl-terminated PDMS oligomer (Mn=1528)

An interesting possible alternative route, which has been reported by Burns and co-workers [38], is shown in Scheme 4. The route takes advantage of the availability of silanol functional fluids of about 1–4000 Mn, which are basically silicone oils. Under proper reaction conditions, the silanol fluid can be end-

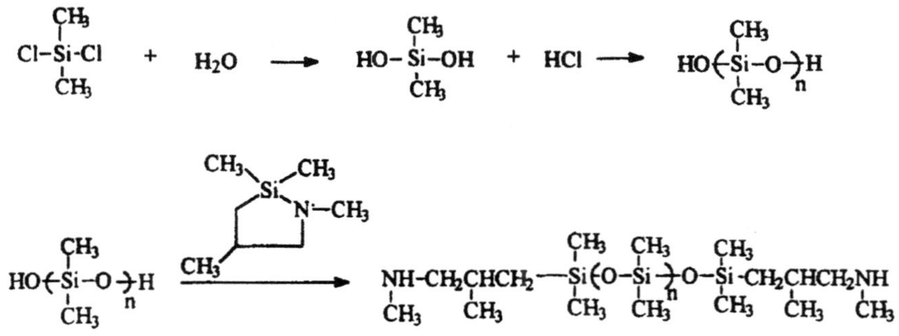

Scheme 4. A possible route for the synthesis of secondary aminoalkyl-terminated polydimethylsiloxane [38]

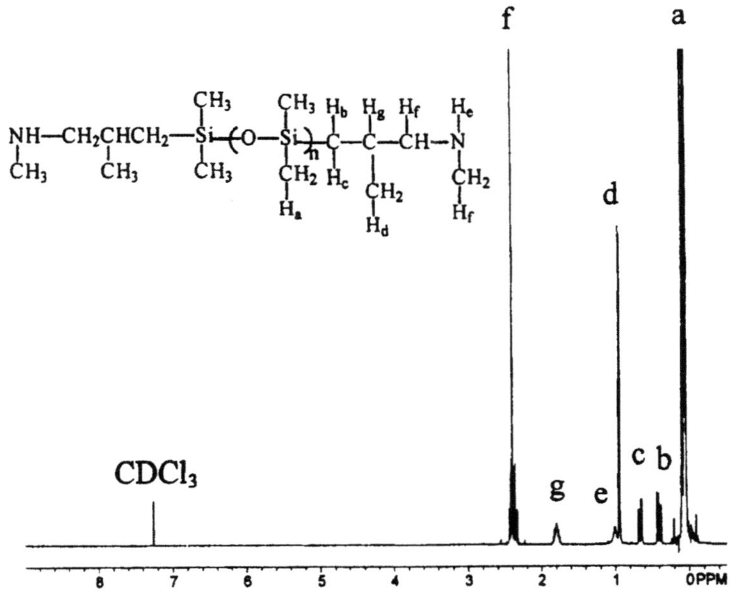

Fig. 5. 400 MHz ^{1}H NMR spectrum of a secondary aminoalkyl-terminated PDMS (Mn= 1235)

capped with cyclics of the type indicated to produce a secondary aminoalkyl functional siloxane. A proton NMR of the secondary aminoalkyl-terminated material is provided in Fig. 5.

The aliphatic n-propyl unit in the siloxane oligomer has been considered to be a source of potential instability and accordingly there has been considerable interest in aromatic aminofunctional polydimethylsiloxane oligomers. Some of the early efforts [39] generated oligomers by reacting the chloropropyl group with aminophenol under nucleophilic displacement conditions to synthesize the aromatic amine functionality. However, this approach has the drawback that the

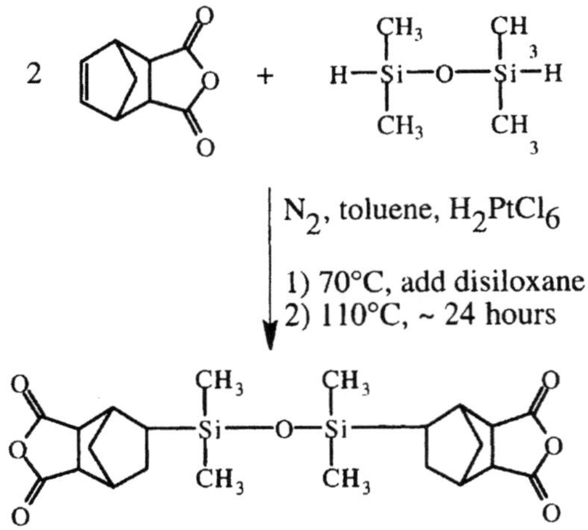

Scheme 5. Synthesis of an anhydride-terminated dimethylsiloxane dimer [44]

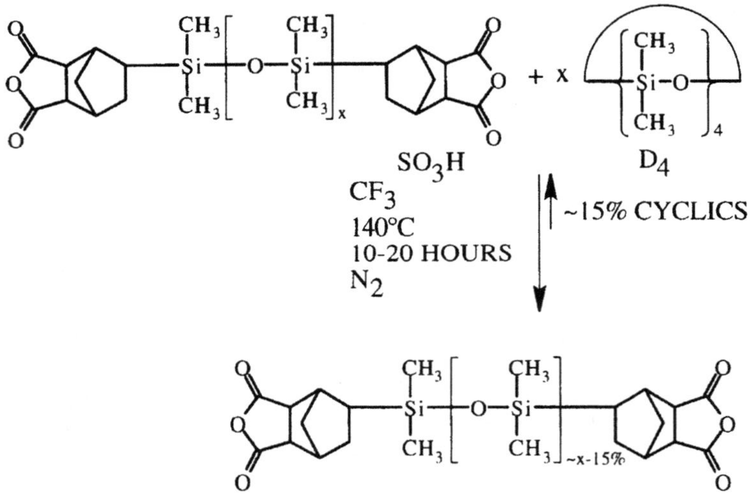

Scheme 6. Acid-catalyzed equilibration synthesis of anhydride-terminated polydimethylsiloxane oligomers [44]

aliphatic structure obviously remains. Alternative approaches investigated the preparation of silicon–oxygen phenyl groups, which, although interesting, suffered from the well-known hydrolysis problem. Later, Riffle et al. [40] synthesized an α,ω-arylamine functionalized polydimethylsiloxane via first preparing the appropriate silicon–aryl bond. The syntheses of the copolymers were also studied by this group and additional work has recently been published by Sysel [41]. These copolymers were prepared via transimidization using approaches that had been developed earlier [42]. There was some improvement in the thermooxidative stability in comparison to polymers based upon aminopropyl or the α,ω arylamine oligomers produced earlier. However, hydrolysis problems, again, might be anticipated for one of the Si–O–AR–NH_2 systems. Quantitative conversion of the aliphatic amine group into the corresponding imide during the copolymerization process itself appears to be quite helpful in enhancing the stability.

An alternate approach for the synthesis of functional oligomers has been to generate a anhydride functionality. The synthesis of the anhydride functionality was investigated by Keohan and Hallgren [43]. The same materials were synthesized by Smith et al. [44], utilizing the procedures described in Scheme 5. The equilibration of the anhydride functional dimer with D_4 was conducted under acid-catalyzed equilibration conditions, as illustrated by Smith et al. in Scheme 6. Earlier, Keohan used a more practical sulfuric acid/SO_3 catalyst [30]. Use of an acid catalyst was necessary to maintain the integrity of the siloxane oligomer anhydride end-groups.

2.2 Copolymer Formation

Segmented copolymers with polyimides were produced by Keohan [43] and also by Smith [44], who also utilized the same coupling reaction with other amino-terminated engineering thermoplastics (ETP) to produce perfectly alternating block sequences. The generalized scheme for the reaction of the anhydride functional oligomer with amine-terminated polyimides or other ETP's is provided in Scheme 7.

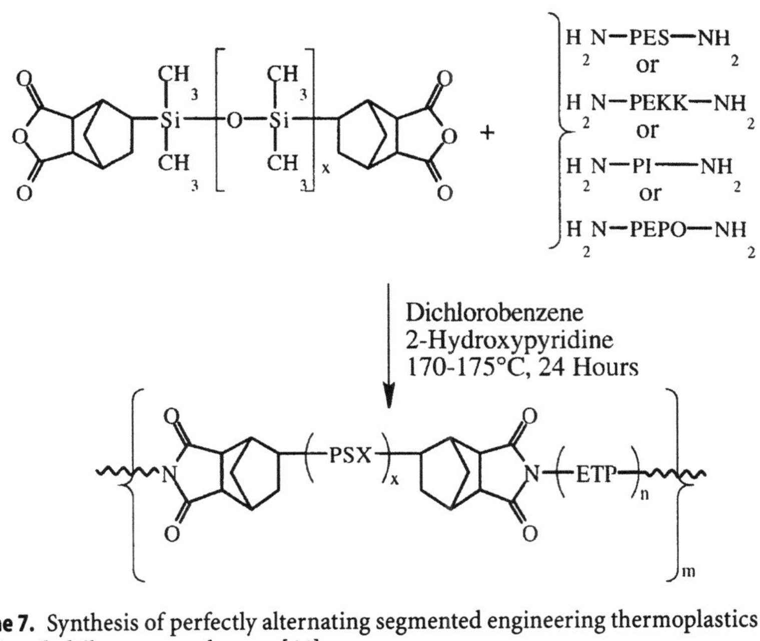

Scheme 7. Synthesis of perfectly alternating segmented engineering thermoplastics (ETP)-polydimethylsiloxane copolymers [44]

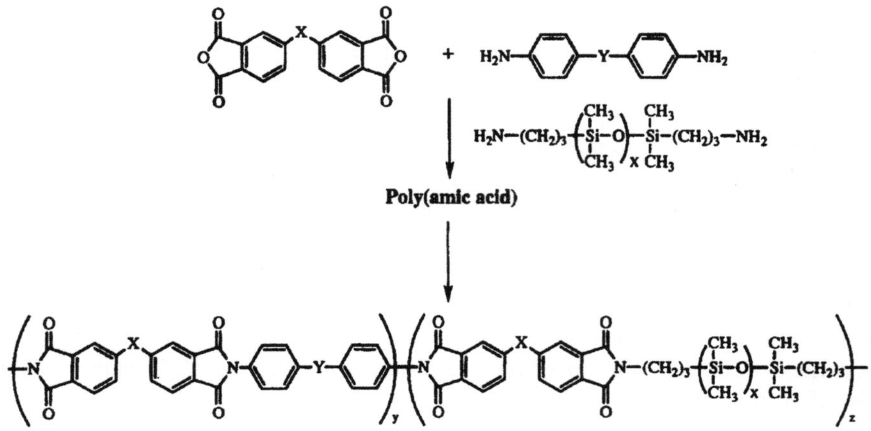

Scheme 8. Synthesis of randomly segmented polyimide-polydimethylsiloxane copolymers [52]

The second, and more common, approach for the synthesis of siloxane segmented copolymers (regardless of the nature of the functional siloxane oligomer) is to prepare what could be called a randomly segmented polyimide polydimethylsiloxane copolymer. A generalized procedure is provided in Scheme 8. The differences in the two systems is subtle, but the sequence length of the imide block is a function of composition in the second approach and this can alter the T_g, etc. The aliphatic aminopropyl oligomer may be first coupled with a dianhydride to enhance the mutual solubility, a fact emphasized by Summers et al. [45], and also in several papers by Arnold et al. [46–50]. The utilization of cosolvents, such as THF/DMAC, was also found to be advantageous and critical to maintain the homogeneous nature of the reactants until the non-polar siloxane oligomer was quantitatively coupled to the amic acid precursor. The imidization second step can usually be done at elevated temperatures, e.g. 180–190 °C, for soluble polyimides under homogeneous conditions. The kinetics and mechanisms of these imidization processes for homopolymers have been discussed by Kim et al. [51] and are equally appropriate for the imide-siloxane copolymer systems, providing, again, that the polyimide phase remains homogeneous.

A second approach for preparation of the segmented copolymers is to first prepare a functionalized polyimide oligomer that can be utilized in a second transimidization step. This general procedure for homopolymers was reported by Takekoshi [5], and adapted by Rogers and co-workers [52], to produce either pyridine or pyrimidine functionalized oligomers. The pyridine or pyrimidine functionalized oligomers can be easily displaced via transimidization with the aminopropyl functional polydimethylsiloxane, under mild reaction conditions to afford perfectly alternating segmented systems, as outlined in Scheme 9.

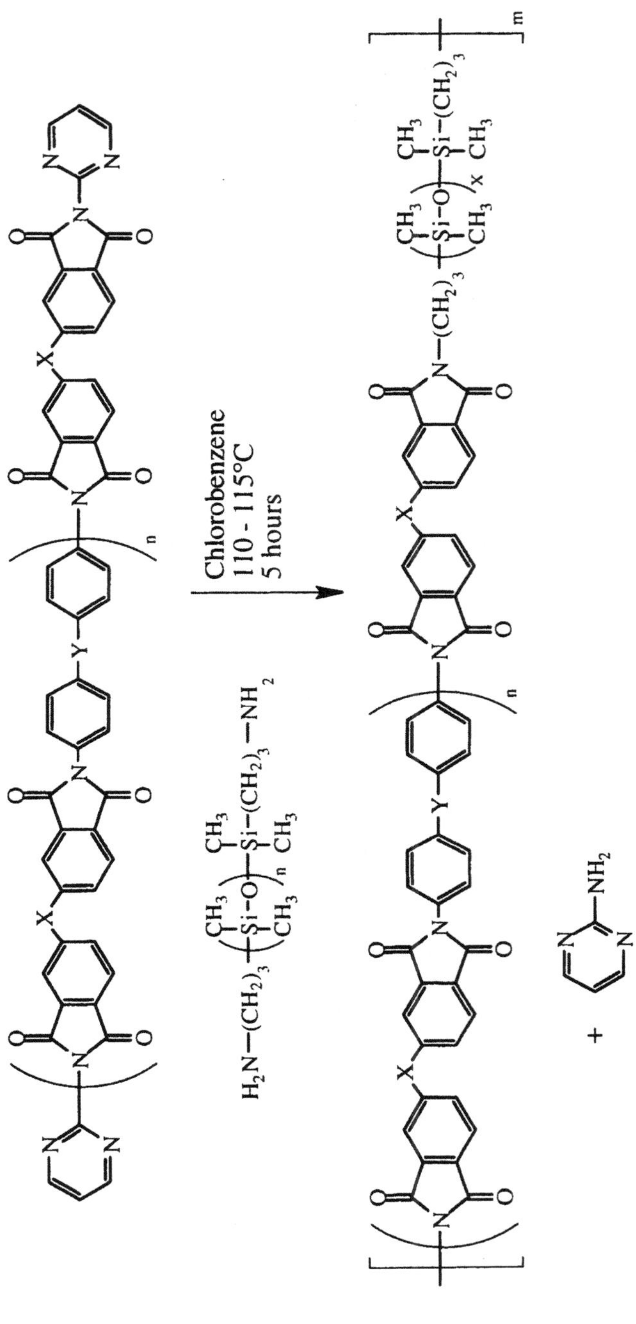

Scheme 9. Synthesis of perfectly alternating segmented imide-siloxane copolymers [52]

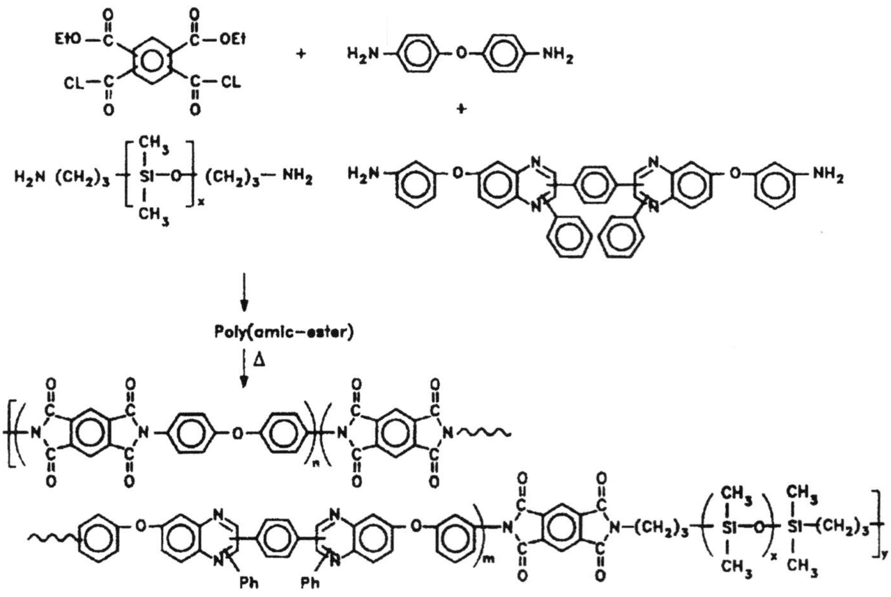

Scheme 10. Synthesis of segmented copolymers via amic ester precursors [53]

A third approach for the synthesis of high performance polyimides has been developed at the IBM laboratories and is referred to as the "amic ester route". This was recently utilized by Hedrick, Volksen et al. [53] to prepare high performance siloxane-modified imides and is illustrated in Scheme 10. The utilization of the ester acid intermediate allows for the simple purification at the intermediate amide ester stage, without the usual problems of time-dependent hydrolysis when a pendant carboxylic acid is present. Bowens et al. [54] have demonstrated that amic acid amine salts can produce environmentally attractive stable aqueous dispersions, which can be successfully melt imidized to afford homo or segmented copolymers. Furukawa et al. [55] have prepared novel thermosetting polysiloxane block polyimides wherein the vinyl functionality has been introduced into the siloxane segment. The resulting materials were soluble and could be crosslinked to produce insolubility by hydrosilation chemistry. It was noted that the modulus increased twofold after crosslinking and that the thermal coefficient of expansion was altered above the T_g, but was fairly similar to that of the uncrosslinked system below the T_g of the polyimide.

3 Characterization

3.1 Molecular Weight and Molecular Weight Distributions

In principle, the molecular weights and molecular weight distributions of the soluble polyimide polydimethyl siloxane copolymers could be established by most of the known techniques, including osmometry, light scattering, size exclusion chromatography, etc. Up to this point, most have relied on the measurement of intrinsic viscosities as an empirical method for estimating molecular weights. Szesztay and Ghadir reported on GPC investigations of polyimide-siloxane copolymers [56]. They used calibrated ultra-styragel columns using the classical polystyrene calibrations. They concluded that the technique provided a useful approximation for samples of molar masses higher than 1,000 g/mol [40]. Arnold and co-workers were also able to establish correlations between intrinsic viscosity and peak elution volume [46–50]. However, they did not carefully utilize calibration procedures. More recently, Konas et al. [57,58] were able to establish universal calibration for polyimides and emphasized the need to introduce additives, such as lithium bromide or P_2O_5, to eliminate the apparent absorption effects of the polyimide solutions in the commonly used solvent NMP. When this was taken into account, the homopolymers agreed very well with the universal hydrodynamic volume calibration procedure. Most likely these approaches would provide a good estimation for the segmented siloxane copolymers as well, but this has yet to be demonstrated. End-group analyses have been well utilized to characterize the siloxane oligomers, as was mentioned earlier. In addition, end functionalization (e.g. with a *tert*-butyl group) of the polyimide homopolymer itself provides a useful NMR estimate of the number-average molecular weight, up to values of as high as 40,000 M_N. This is an additional method for correlating the spectroscopic and chromatographic methods. Most investigators have been primarily concerned with correlating the increase in intrinsic viscosity with the ability to form strong coherent films from the copolymers. The development of transparency has been used as a qualitative guide to enable confidence that the siloxane has been essentially quantitatively incorporated into the copolymer. Small amounts of residual cyclics may be either present initially or possibly generated throughout the reaction or during processing, and the observations are usually that the observed siloxane content by proton NMR spectroscopy is slightly lower (usually within experimental error) than the calculated value. One precaution is to ensure that the deuterated solvents used for the NMR measurements are sufficiently dried to prevent any apparent micellization of the block copolymers. Solvents containing detectable concentrations of water have, in some cases, been observed to produce lower than the actual siloxane content values [52].

3.2
Copolymer Composition

The concentration of the siloxane introduced into the copolymer is most easily determined by proton NMR for soluble systems, as briefly referred to in Sect. 3.1. This was demonstrated by Summers et al. [45] and Arnold and coworkers [46–50], as well as Rogers et al. [52]. Others have routinely conducted these experiments to establish the copolymer composition. Infrared spectroscopy has been particularly useful for demonstrating the transformation of the amic acid, which is often an intermediate for the final imide form. The assignments have been noted in many of the polyimide reviews referred to earlier [1–8]. In addition, it is useful to conduct an elemental analysis for silicon as complementary proof of copolymer composition. Solid-state NMR can be used even for intractable polyimide systems to provide a good estimate of the copolymer composition.

3.3
Thermal Behavior

The thermal behavior of segmented polysiloxane-polyimides is analogous in many ways to other block copolymer systems described earlier [10]. Providing that one exceeds the block length necessary to produce a microphase separation, one can usually observe the glass transition temperature of the polydimethylsiloxane well below −100 °C by methods such as differential scanning calorimetry or dynamic mechanical behavior. The upper T_gs can usually also be detected. The dynamic mechanical loss peaks are generally more sensitive than the calorimetric methods, but usually both can provide a relatively good definition of the transition area. The upper transitions due to the polyimides may be considerably depressed relative to the value expected for the high molecular weight homopolymer. This has been interpreted by some as being due to partially miscible phase behavior between the hard segment and the soft siloxane block. However, alternatively, the number-average sequence length of the hard block is expected to be shorter than it would be in the homopolymer – at least when randomly segmented copolymers are prepared. Therefore, the depression in the upper T_g may also be associated with the shorter hard segment length in this case. This issue was further addressed by Rogers and co-workers [52] who prepared perfectly alternating segmented copolymers via transimidization. This allowed the chemistry (which was described earlier) for the coupling of preformed blocks of well-defined number-average size. The resulting transition temperatures were found to be a strong function of the original segment length of the siloxane functionalized oligomer, as demonstrated in Fig. 6. The segmented copolymers of the 4100 g/mol hard segment displayed T_gs a little above 200 °C, as shown in Fig. 6. Note that the siloxane transition appears to be unchanged as a function of composition. However, as anticipated, the modulus between the two transitions is much higher for the 20% copolymer than it is for the 50% siloxane

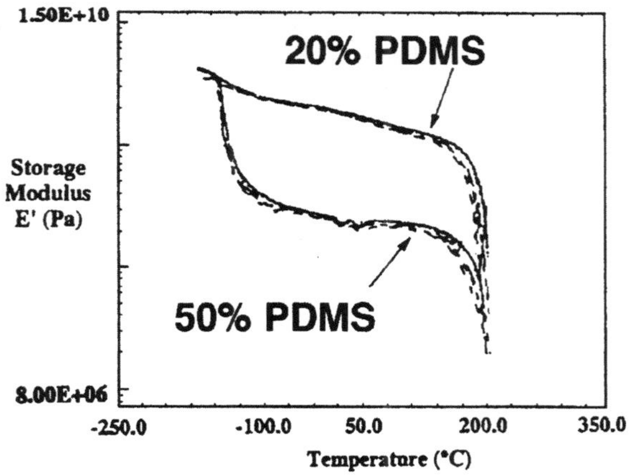

Fig. 6. Influence of composition on the modulus-temperature behavior of perfectly alternating block copolymers [52]

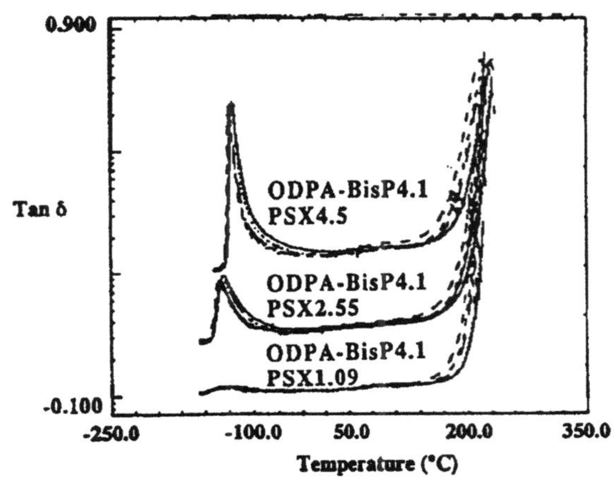

Fig. 7. Influence of composition on the damping behavior of perfectly alternating block copolymers [52]

system. Thus, the 20% copolymer films are tough, ductile transparent plastics, whereas the 50% composition approaches thermoplastic elastomer behavior. The transitions are further confirmed by the analogous damping peak, the magnitude of which is observed to increase with the length and composition of the copolymer, as expected (Fig. 7). Thus, increasing the siloxane number-average molecular weight and hence composition from about 1100 to 4500 gm/mole can be observed to increase dramatically the damping peak at −100 °C. The upper transition temperature for the polyimide is relatively unaffected by this compositional change and suggests that, indeed, the value is essentially defined by the constant length of the polyimide, which was based upon oxydiphthalic anhydride (ODPA) and bis-aniline-P (Bis-P) chemistry as shown below. The influence of composition on the mechanical behavior is discussed in the next section.

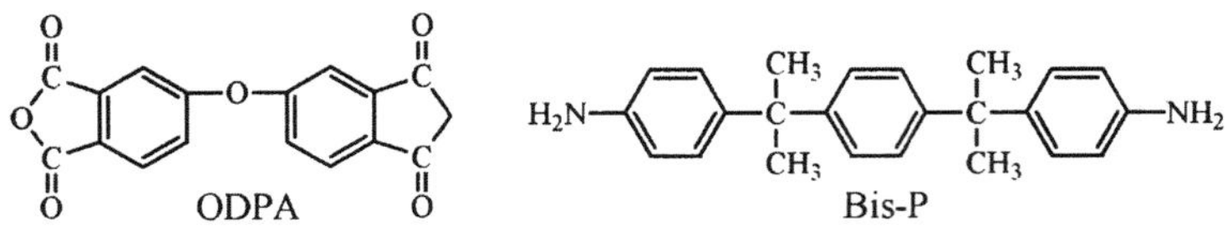

3.4 Structure-Property Relationships

3.4.1 *Morphology and Mechanical Behavior*

The morphological behavior has been investigated by a number of workers. York et al. [59] investigated the influence of composition on the transmission electron microscopy (TEM) behavior and were able to develop correlations between dispersed, co-continuous and inverted types of morphologies as largely outlined earlier for hydrocarbon-based systems [10]. The morphology observed was influenced by casting solvents, as well as the processing techniques. The morphology of the perfectly alternating systems prepared by Rogers [52] was examined by transmission electron microscopy (TEM), as shown in Fig. 8. One can distinguish the siloxane phase due to its higher electron density relative to the poly-

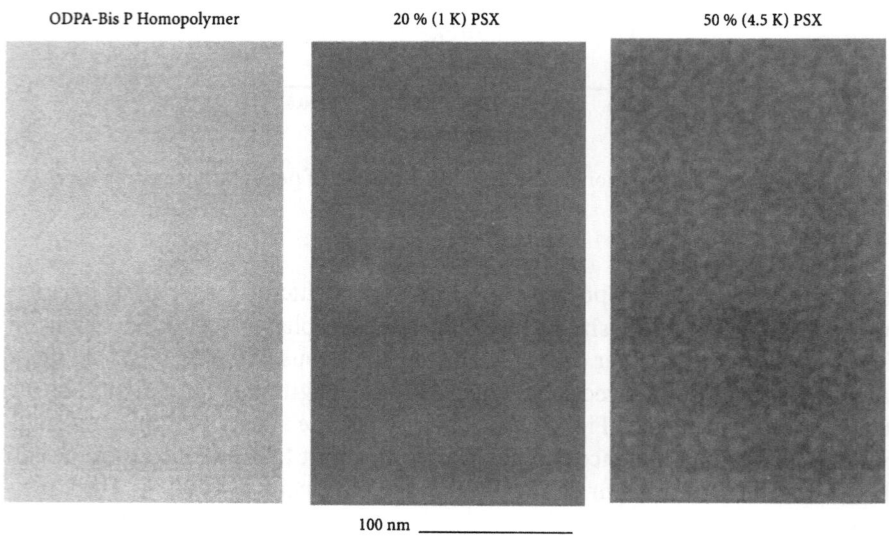

Fig. 8. TEM studies of the effect of polydimethylsiloxane concentration and molecular weight on microphase development

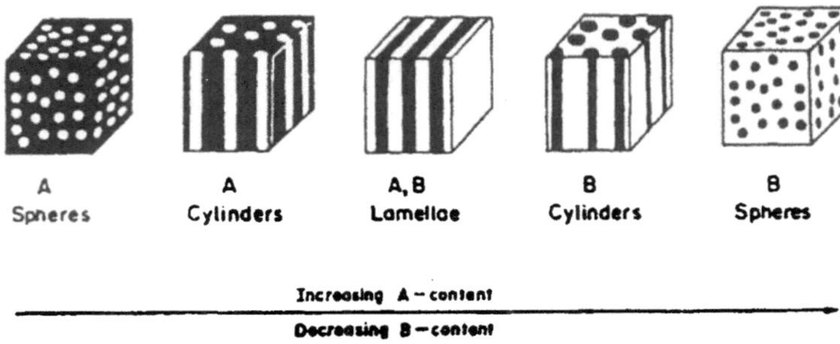

Fig. 9. Classical microphase separated morphologies [10]

imide. Thus, the amorphous polyimide homopolymer, (ODPA-BisP) is essentially featureless. At 20 wt% siloxane in a copolymer based on a 1000 g/mol molecular weight polydimethylsiloxane, a very fine microphase separated structure is observed. Upon increasing the concentration and molecular weight of the siloxane block to about 50 wt% and 4500 g/mol, respectively, a quite well-defined co-continuous type morphology is observed. This was consistent with earlier studies on block copolymers, e.g. spheres, rods, lamellae, inverted rods, inverted spheres, etc. (Fig. 9) and can be correlated with the mechanical behavior. Thus, Arnold et al. in several papers [46–50] investigated the stress-strain behavior at room temperature of a polyimide based on BTDA and 3,3'-diaminodiphenyl sulfone (DDS) (shown below) based segmented copolymers and was able to establish structure-property relations, as illustrated in Fig. 10.

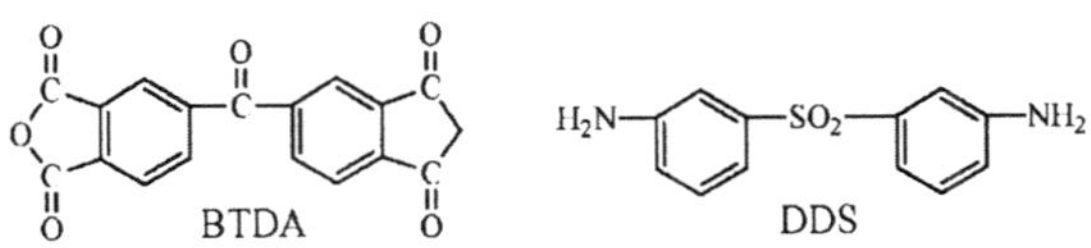

Thus, at 70% by weight of siloxane, the material was capable of more than 500% elongation prior to failure, while showing ultimate stress at failure approaching 20 MPa. Upon increasing the concentration of the imide to 60% (40% siloxane), the elongation at break was reduced to less than 100%, but the failure stress had increased to about 40 MPa, e.g. a factor of 2. Further increases in the imide concentration to 80 and 90%, (e.g. 20 and 10% siloxane), respectively, produced stress-strain curves at room temperature where the initial moduli was nearly as high as the control polyimide. This was interpreted as being due to the fact that the imide was now the continuous phase and the siloxane was a dispersed phase. There was an increase in the ductility or toughness of the films as observed by a more significant yielding for the siloxane-modified systems. Saraf, Feger and Cohen [60] reported a detailed structural characterization of a commercially provided imide-siloxane copolymer. They were able to establish a number of important features using classical scattering techniques, such as small angle X-ray scattering (SAXS), and proposed that a three-phase model might be more ap-

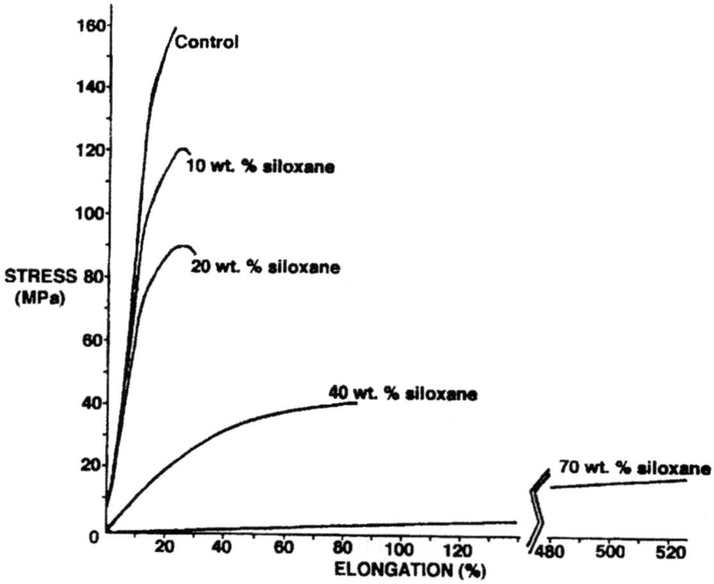

Fig. 10. Stress-strain behavior of BTDA-m-DDS-based polydimethylsiloxane-polyimide segmented copolymers [46]

propriate for their molecular structure. Unfortunately, neither the chemistry nor block lengths were indicated in this paper, so that further interpretation is quite difficult. Furukawa, Yamada, and co-workers have also reported on the surface and morphological characterization of polysiloxane block polyimides [61–63]. Their dynamic mechanical behavior show that the storage modulus vs. temperature was consistent with that reported by Rogers et al. [52] and the tensile modulus was significantly decreased from the control polyimide as one prepared from 15, 30 and finally 50 wt% siloxanes.

3.4.2
Adhesives and Adhesion Science

The potential widespread utilization of polyimides for structural adhesives, matrix resins, electronic packaging, and many other applications is sometimes limited because of processing issues. It was recognized some time ago by St. Clair and co-workers [16–17] and Bott et al. [64–66] that segmented polyimide-siloxane copolymers have great potential as structural adhesives for metals, e.g. titanium. Materials were prepared to further investigate this potential either by bulk casting of the amic acid precursors or by solution imidization by Summers, Arnold and co-workers [46–50]. Significantly improved properties were demonstrated through the use of thermal solution imidization for both certain homopolyimides, and especially for segmented polyimide-siloxane copolymers. The

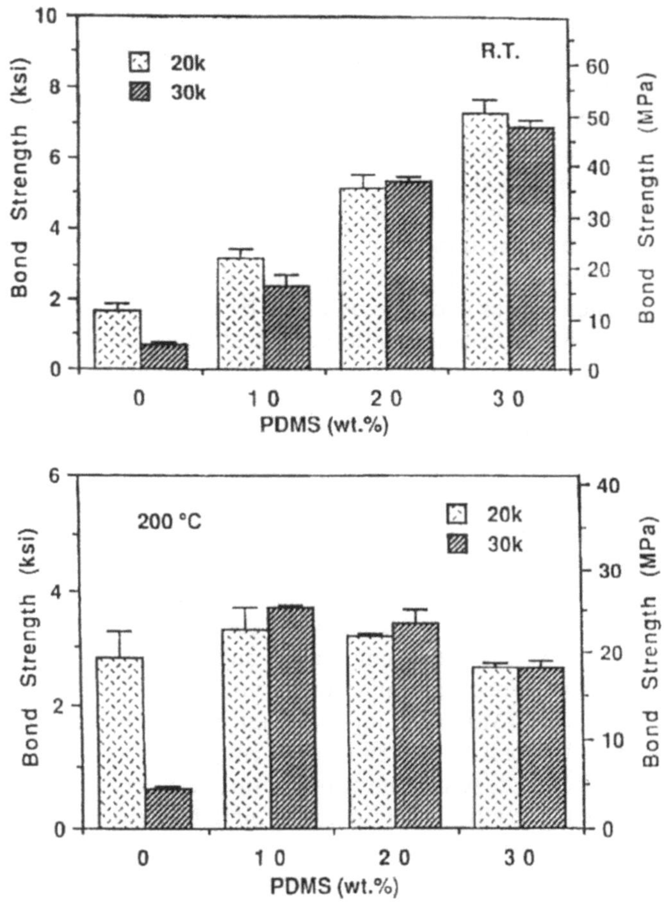

Fig. 11. Effect of siloxane incorporation on adhesive bond strength at rt and 200 °C

enhancement of solubility and processability and moisture reduction was attributed to the siloxane phase. Small amounts (e.g. 10–20%) of the siloxane did not detract from the lap shear strength of these materials to titanium and improved the resistance to hot, wet environments. The added durability was interpreted as being due to the hydrophobic siloxane segments, and their ability to reduce the uptake of water at the critical interface between the adhesives and the adherends. Bott et al. [64–66] discussed procedures for generating high bond strengths as well as the durability for structural adhesives. The observations were continued and expanded in a series of papers by Yoon et al.[67–73]. The adhesive performance of thermoplastic imide-siloxane to polymers was investigated, not only to titanium but also to polyarylene ether ketone (PEEK) graphite composites. The adhesive bond performance of the PEEK graphite composites with melt fabricated films derived from either a 30% imide-siloxane copolymer or a thermoplastic polyimide was studied as a function of surface modifications, which included grit blasting and gas etching treatments with oxygen, ammonia, nitrogen and argon. The adhesive bond strength was evaluated by single lap shear specimens and it was observed that greatly improved single lap shear strength could be obtained from these systems.

Further studies by Yoon et al. [68] showed that there was a relationship between the siloxane phase dimensions, concentrations, and adhesion strength, particularly as a function of temperature. It was demonstrated that outstanding lap shear adhesion values could be obtained by using segmented copolymers containing 20–30 wt% of siloxane, but that this strength value decreased somewhat with elevated temperatures. In contrast, the materials which contained only 10 wt% of siloxane had excellent adhesive properties over a wide temperature range (Fig. 11). A model was developed to explain these phenomena in terms of a balance between chain rigidity and ductility.

Yoon et al. [68] used thermoplastically processable fully cyclized BTDA, DDS based polyimide homopolymers and the segmented copolymers which were designed to have non-reactive phthalimide end-groups. This degree of molecular control allowed excellent melt laminated single lap shear samples to be prepared from surface treated titanium 6/4 adherends. Tests were conducted from room temperature to 250 °C. The room temperature adhesive bonds were best for the 30 wt% siloxane and were optimum for the homopolymer control when it had a total molecular weight (Mn) of about 20,000 g/mol. This work emphasized the importance of melt viscosity as well as segmental mobility of the copolymer. Some of the lap shear values at room temperature were very high, 50 MPa (7000 psi), again suggesting that the combination of melt flow and bond consolidation were both critical parameters. The modulus temperature behavior was assessed by a thermomechanical penetration measurement, which indicated the practical flow temperature. The segmented copolymers behavior was influenced not only by the T_g, but also by the molecular weight of the hard segment.

The behavior of the adhesives as a function of temperature by TMA measurements is illustrated in Fig. 12. Here the softening points for the UEC designation, which represent uncontrolled molecular weight and the other controlled molecular weights illustrate that the glass temperatures are relatively similar for all of the polymers, but that the elastic entanglement region above the T_g is very much influenced by the molecular weight. This correlates very well with the observations of the adhesive bond strengths at room temperature and also at 200 °C.

The general area of adhesion and adhesion science has been one of the most important application areas identified thus far in a variety of applications for the subject imide-siloxane copolymers [74–82]. Many references in the patent literature [83–90] have been found that relate to the utilization of the copolymers in various important applications such as die attach adhesives, encapsulants, interphase agents, etc. It would appear that the combination of very good thermal stability, coupled with the adhesive bonding capabilities provided by the siloxane mobile segment, has attracted a great deal of attention and will be the

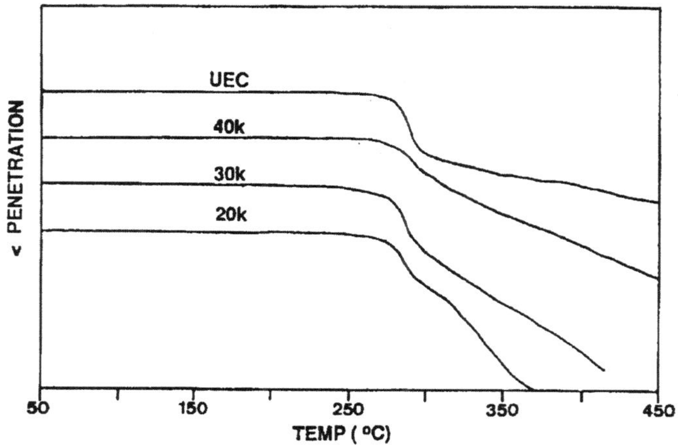

Fig. 12. Influence of Molecular Weight and End Capping on TMA behavior of polyimide homopolymers (in air, 10 °C/min)

subject of future development. Unfortunately, many critical molecular details are omitted in the patents and hence it is not possible to draw more quantitative conclusions.

3.4.3
Low Stress Thin Film Dielectrics

The utilization of polyimide thin film dielectrics in computer applications is well illustrated [1–8]. The siloxane-imide block copolymers are also of interest in applications that can accept moderately less thermal stability, but perhaps need improved flexibility, adhesion, and toughness. The issue of stress buildup in thin film dielectric films or coatings has been reviewed by Farris and co-workers [91] and will not be discussed here except to indicate that it can lead to premature failure and two approaches have been used to solve the problem. One is to design low thermal coefficient expansion (CTE) polyimides, which are intended to match the substrate. These materials are often based on liquid crystalline or rigid rod polyimides [1–8], but they are anisotropic and this introduces additional, unwanted problems, such as an anisotropic dielectric constant. It is well known that reducing the modulus with a high concentration of rubber-like material in the composition can reduce the stress, but will also usually reduce the thermal stability and modulus below tolerable levels. Recently, Hedrick et al. [53] have demonstrated that relatively small amounts, e.g. 20% or less, of phase-separated polydimethylsiloxane copolymers can be utilized to significantly lower the stress in coated films, compared to the control polyimide. This is illustrated in Fig. 13, where the apparently zero stress system corresponds to a 20 wt% segment of siloxane of about 5000 Mn. The moderately lower stress sample utilized

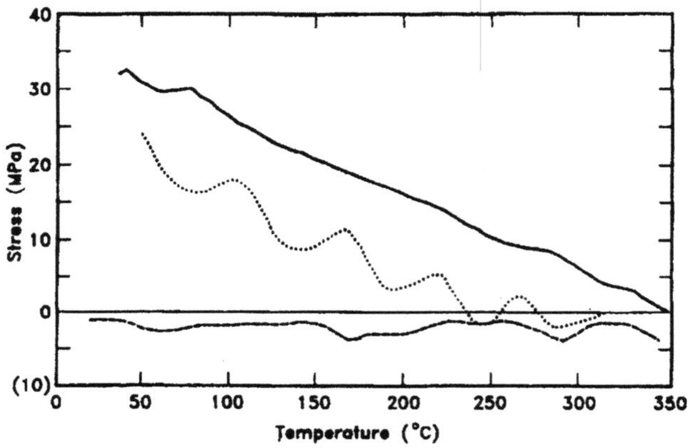

Fig. 13. Reduction of residual stress vs. temperature plots for copolymers 3 (.....) (1000 Mn) and 5 (–) (5400 Mn) vs. PMDA/ODA polyimide control [53]

a 1000 Mw oligomer, which was also significantly better than the polyimide. Furukawa has also reported stress relaxation behavior of segmented copolymers [92].

3.4.4
Fire Resistance

Methodologies to improve the fire resistance of organic materials, such as rubbers, plastics and fibers, is of continuing importance. Traditional methods include the utilization of additives such as brominated ethers, antimony oxides, etc. The incorporation of inorganic elements such as silicon or phosphorus into the chemical backbone is of considerable interest. It has been demonstrated that the silicon-containing polyimides degrade in air to produce a char. A generalized view of char formation is provided in Fig. 14. The development of a char occurs much more readily in an inert atmosphere such as nitrogen than it does in air or oxygen for organic materials. This subject has been reviewed recently in a book edited by Nelson [93]. In the presence of air, heteroatoms such as silicon or phosphorus are of considerable interest since the degradative reactions can produce silicate- or phosphate-like derivatives. This is illustrated in Fig. 15 where the thermogravimetric analysis was measured dynamically in air for a series of segmented siloxane-imide copolymers that varied in siloxane concentration from 0 to 60 wt% . The resulting curves do show less thermal stability as a function of the siloxane content in terms of their initial weight loss. This could be attributed to a number of degradative mechanisms, which have been investigated [94] in the past. However, at very high temperatures in air of about 700 °C, this series of imide-siloxanes was noted to show a significantly increasing char concentration. When the residue was analyzed, it could be demonstrated by XPS that the material contained silicate bonds and that silicon–carbon bonds had

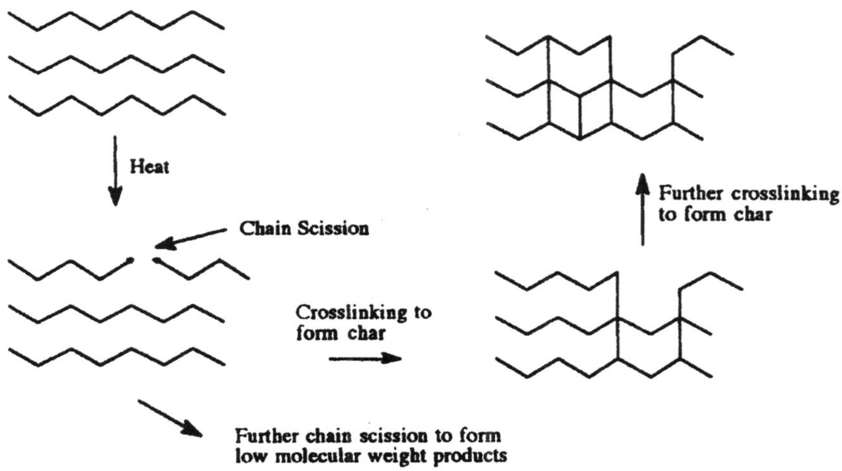

Fig. 14. A mechanism for char formation

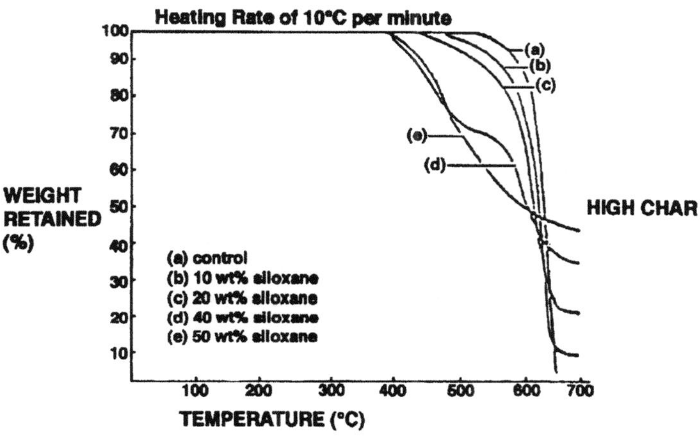

Fig. 15. Thermal gravimetric analysis in an air atmosphere of poly(siloxane imide)-segmented copolymers of various siloxane concentrations, indicating char yield proportional to siloxane content

been thermally broken at these high temperatures as illustrated in Fig. 16. This is in contrast to the most common method of degradation for polydimethylsiloxane homopolymers, which involves reversion at much lower temperatures. One may speculate that tying up the ends of the siloxane chain allow it to (at least briefly) pass through temperature intervals that might otherwise lead to reversion and allows partial conversion to silicate-like structures in an air atmosphere. Independent observations, e.g. Bunsen burner tests or more quantitative

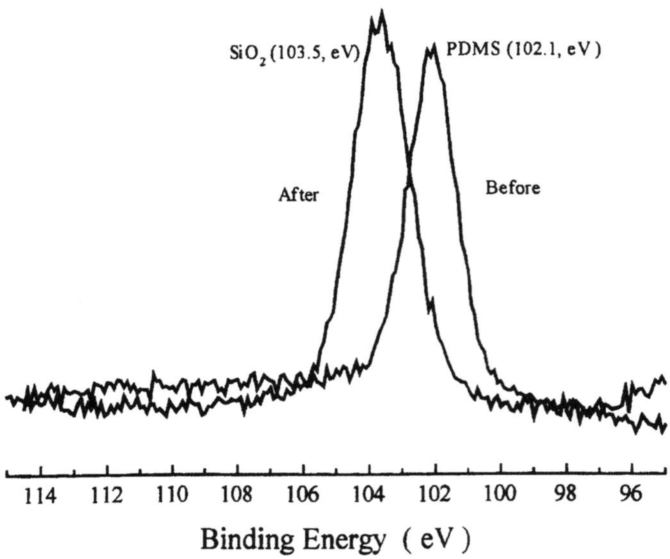

Fig. 16. XPS Analysis of PDMS containing copolymer before and after exposure to 700 °C

cone calorimetry, indicate that the generation of the silicate char interrupts the burning process and does so desirably with only limited production of smoke, in contrast to the behavior of many aromatic polymers. Polyimide-siloxane copolymers were also prepared wherein the imide portion was derived from a phosphine oxide comonomer [95]. These hybrid materials also showed excellent fire resistance. This area of research will no doubt continue as the search for more efficient and safe fire resistance materials continues.

3.4.5
Gas Separation Membranes

The permeability of polymeric films is of great interest at both the low values in food packaging, and for the high permeability levels for selected purification of, for example, oxygen/nitrogen or carbon dioxide/methane streams. It has been known for a long time that polydimethylsiloxane has a very high permeability coefficient and this has been investigated and utilized. However, the selectivity of the siloxane polymers is relatively low. For example, homopolymers are known to have an approximate O_2/N_2 separation ratio of about 2.0. A number of workers have investigated the possibility that imide-siloxane block copolymers might show some interesting behavior. Indeed, the imide segment can certainly provide mechanical strength to the otherwise weak siloxane phase, as illustrated earlier. Correspondingly, the siloxane soft rubber phase can also toughen the imide block. The correlation between the microphase separation and the basic

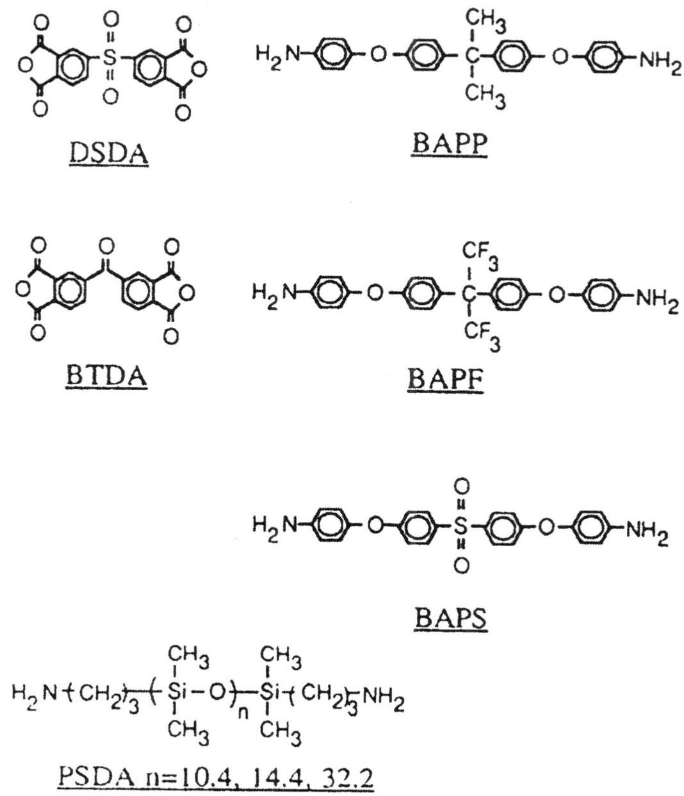

Scheme 11. Monomers and oligomer for permselective membranes [97–99]

morphological structures has been of continuing interest. Stern and co-workers [96] were able to prepare good films and carefully study their permeability. Unfortunately the compositional range available to them did not have any interesting selectivity. However, others such as Yamada and co-workers [97–99] have studied a range of membranes, based on the building blocks shown in Scheme 11. The copolymers were investigated and prepared in high molecular weight and their permeability coefficients were measured. The gas separation properties of polyimides with various aromatic diamines, including siloxane oligomeric diamines from 10–40 wt%, wherein the degree of polymerization was estimated to be slightly more than 10, are provided in Table 1. The values show that as a function of the various aromatic amines (with a common dianhydride, BTDA), a range of permeabilities and separation ratios can be obtained. Note that the highest siloxane concentration of about 40% only has an O_2/N_2 ratio slightly higher than the control. However, other mechanical strength issues might well be quite favorable for these materials. The influence of composition was also investigated by Mecham and coworkers [100–102] who utilized the per-

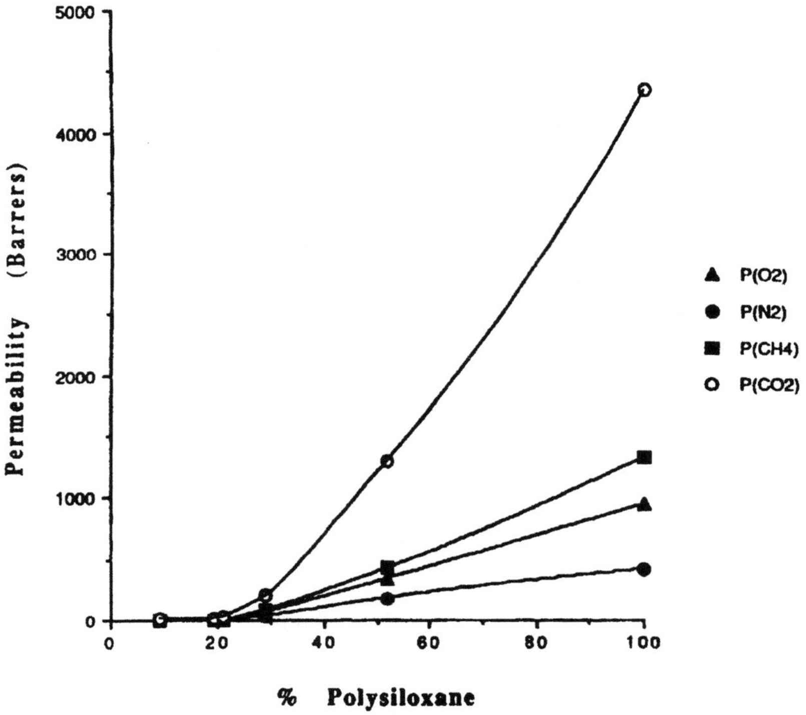

Fig. 17. The effect of PDMS concentration on the permeability coefficients of ODPA-Bis P/PDMS perfectly alternating block copolymers for O_2, N_2, CH_4, and CO_2

Table 1. Gas separation properties of polyimide-polydimethylsiloxane segmented copolymers as a function of the aromatic amine used for the hard segment [99]

Aromatic Diamine	Silicone Diamine Content (wt%)	$^{P}O_2$	$^{P}N_2$	$^{P}O_2/^{P}N_2$	$^{P}CO_2$	^{P}CH	$^{P}CO_2/^{P}CH_4$
BAPP	-	0.570	0.100	5.7	2.12	0.060	35.3
	10	0.902	0.181	5.0	3.27	0.096	34.0
	20	4.34	0.994	4.4	22.8	1.74	13.1
BAPS	-	0.287	0.044	6.5	1.14	0.027	43.0
	10	0.452	0.106	4.3	1.72	0.043	40.0
	20	1.07	0.260	4.1	4.72	0.140	33.7
BAPF	-	1.03	0.162	6.4	4.18	0.094	44.4
	10	1.83	0.294	6.2	7.31	0.213	34.3
	20	5.65	1.49	3.8	25.7	2.18	11.8
	40	55.5	25.6	2.2	335	67.0	5.0

Tetracarboxylic dianhydride: BTDA. Units: P[$10^{10} \times cm^3$ (STP) $\times$ cm/(s $\times$ $cm^2 \times$ cmHg)]. Temperature: 25° C. Sequence length of silicone diamine: n=10.4

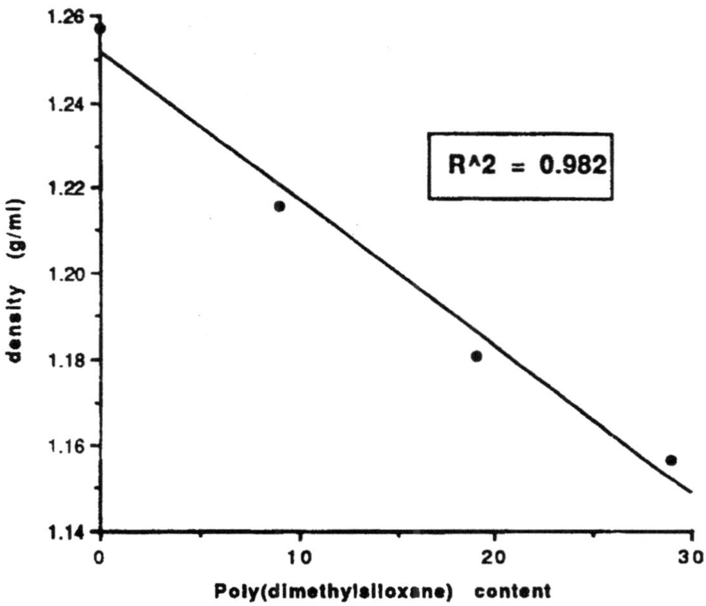

Fig. 18. Density vs. PDMS content for a homopolyimide and the polyimide/poly(dimethylsiloxane) perfectly alternating block copolymers with the 11,000 g/mol polyimide segments

fectly alternating copolymers of various compositions generated by Rogers et al. [52]. The resulting cast films were investigated as a function of conversion and dramatic differences could again be observed slightly higher than 20 wt% as shown in Fig. 17. Density was observed to be approximately linear over the 0–30 wt% PDMS range (Fig. 18). Many other detailed studies of the gas and liquid membrane behavior of these interesting copolymers have been reported (103–114) and, in general, the microstructural response and bulk characterization are consistent with the multiphase morphology features outlined earlier. This has been illustrated by a number of solid-state structural investigations, especially including those of Spontak et al [115–118].

3.4.6
Aerospace, Electronic Applications, and High Performance Blends

The earlier sections of this review have attempted to review the synthesis and characterization of the segmented copolymer materials as well as to highlight fundamental efforts related to a number of structure-property relationships. Polyimide-siloxane copolymers are high performance materials that have been driven by applications frequently identified with aerospace. However, this section will also include surface phenomena which are critical for certain electronic applications and high performance blends utilizing imide-siloxane copolymers.

These materials might be useful for advanced airplane materials and/or outer space applications, including satellites where either atomic oxygen or high energy radiation is an important mode of degradation. There is a wealth of additional information principally in the form of patents that appear to fall into the current subject area. Although pertinent details are often lacking, a summary of a number of major polysiloxane-polyimide copolymer patents not otherwise referred to is provided in Table 2. They comprise applications including adhesion (covered earlier) but also abrasion, circuit boards, tape automated bonding, prepregs, injection moldable block copolymers, crosslinked systems, and materials for bonding thermoplastics, metals and ceramics.

This section describes efforts to apply the copolymer to new aerospace materials. Much of this initial work has been sponsored by NASA in the United States, who have been interested in developing materials which can withstand the rigors of outer space. One issue is to define structural materials which can resist both high energy radiation and also aggressive oxygen, primarily in the form of atomic oxygen. The siloxane-polyimide copolymers were some of the first materials identified for this purpose, and the properties of interest include toughness, thermal stability and the selective surface microphase separation of polydimethylsiloxane-containing block copolymers and blends. This latter subject has been reviewed by Dwight and co-workers [119], who further developed the idea that if one achieves a microphase separated copolymer in bulk, there will be a driving force for the migration of the lower surface energy component to the air or vacuum interface. Such phenomena have been studied by surface analysis techniques such as contact angle or XPS, as cited earlier [75]. The XPS step profiling of imide-siloxanes is illustrated in Fig. 19. At the low θ or grazing angle

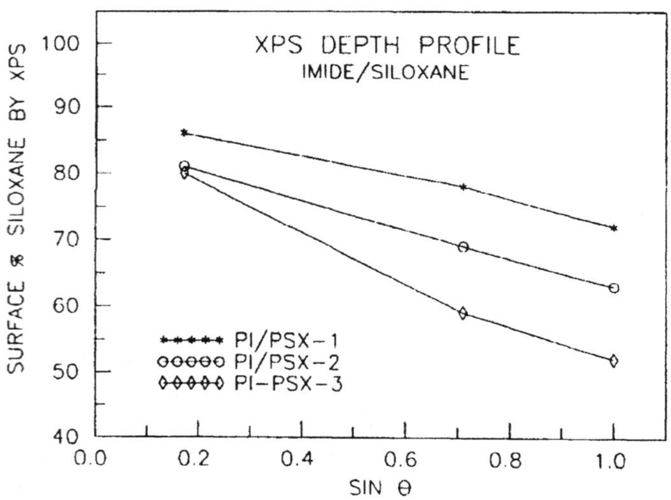

Fig. 19. The XPS step profiling of imide-siloxanes segmented copolymers [119]

Table 2. Polysiloxane-polyimide copolymer patents

Patent No.	Title	Assignee; Inventor	Uses/Advantages	Technical Summary
JP 05009254 (Japanese) 1993	Self-lubricating electrically insulating wires	Fujkura Kk, Japan, S Amano et al.	Insulating coating was improved for Cu wire with good lubricity and abrasion resistance	Polyimide-siloxane was prepared using PMDA, MDA and aminosiloxanes
JP 05125345 (Japanese) 1993	Poly(vinyl butyral) blend adhesive composition for flexible printed circuit boards	Ube Industries, Japan; H Inoe, S Takabayashi, T Muramatsu, T Funakoshi	Formulated resins having good flexibility and heat resistance are useful adhesives for polyimide/Cu foil laminated circuit boards	Maleimide-terminated siloxane-imide block copolymers were formulated with poly(vinyl butyral) and melamine resins to give adhesive compositions.
JP 05197182 (Japanese) 1993	Electrophotographic Photoreceptor with Amorphous Silicon Photoconductive Layer	Fuji Xerox Co. Ltd., Japan; S Yagi et al.	Copolymer useful for surface protective coating; photoreceptor shows stability and gives sharp images	Maleimide-terminated siloxane-imides crosslinked coatings were used to protect photoreceptors.
JP 05306386 (Japanese) 1993	Heat-resistant adhesive composition	Ube Industries, Japan; H Inoe, S Takabayashi, T Muramatsu, T Hirano	Resin formulation useful for bonding Cu foils to polyimide films for electronic applications	Heat-resistant adhesives were formulated to include imide-siloxanes with softening points <300° C. Diaminopolysiloxanes and aromatic amines were used; maleimide termination also utilized.
JP 05311144 (Japanese) 1993	Heat-resistant adhesive compositions containing polyimide-siloxane block copolymers	Ube Industries, Japan; H Inoe, S Takabayashi, T Funakoshi, K Sonoyana	Useful for Cu-clad laminates of printed circuit boards and tape-automated bonding with good flexibility and heat resistance	Formulated compositions containing siloxane-imide block copolymers made from aromatic tetracarboxylic acids and aromatic diamines and aminosiloxanes
EP 0518060 (European) 1992	Polyimide-siloxane extended block copolymers	Occidental Chemical; S Rojstaczer	Useful as coatings and adhesives in microelectronic; have higher T_g than the corresponding non-extended block copolymers while retaining solubility	Siloxane-imide block copolymers were prepared using amino-terminated siloxanes and organic diamines along with pref. aromatic dianhydrides
JP 04220431 (Japanese) 1992	Tough polyimides and prepregs and fiber reinforced plastics	Toray Industries, Japan; M Eno, K Tobukuro	Toughened fiber reinforced prepregs	Polyimide-siloxanes prepared from BTDA and aminosiloxanes with nadic anhydride termination

Table 2. (continue)

Patent No.	Title	Assignee; Inventor	Uses/Advantages	Technical Summary
US 5,028,681 (American) 1991	Novel poly(imide siloxane) block copolymers and process for their preparation	General Electric; EN Peters	Injection moldable block copolymers with high IV and excellent chemical/physical properties. Blends useful for impact modification	Novel siloxane-imide block copolymers and a process for their preparation are covered. The method involves reacting a hydroxy-terminated polyimide oligomer with a siloxane oligomer with dimethylamino, acetyl or chlorine end-groups
US 5,204,399 (American) 1991	Thermally conductive thermoplastic polyimide film die attach adhesives and their preparation	National Starch and Chemical Investment Holding Corp.; R Edelman	Adhesive formulations having excellent die shear strength	A polyimide-siloxane was prepared from aromatic amines, bis aminophenoxybutylsiloxane and aromatic dianhydrides
EP 323142 (European) 1989	Ternary polyether ketone blend wire insulations	Pirelli General PLC, UK; CK Alesbury, RJ Murphy	Formulation shows excellent stress craze resistance, flexibility and flame resistance	Blends of polyarylene ether ketones, polyether imides and poly(imide siloxanes) were coated onto wire for solvent resistance, O index and abrasion resistance. Blends without siloxane-imide copolymer did not meet these properties
FR 2612195 (French) 1988	Thermally stable polymers from bismaleimides, siloxane diamines and optionally maleimide-terminated siloxanes	Rhone-Poulenc Chimie, Fr.; R Barthelemy, Y Camberlin	Heat-resistant maleimide-terminated siloxane	Thermally stable polymers were prepared by co-curing MDA based bismaleimides and maleimide-terminated siloxanes. Michael addition with amino-terminated siloxanes and MDA was also utilized.
JP 62246978 (Japanese) 1987	Heat-resistant adhesives containing siloxanes and imide-containing resins	Shin-Etsu Chemical Industry Co. Ltd., Japan; S Ueno, N Nakanishi, S Hoshida	Adhesives showed good adhesion and flexibility useful in the preparation of flexible laminates	Resin was prepared from the reaction product of bismaleimides, diamines, amino-terminated siloxanes and a bisphenol A cyanate adduct. The adhesive formulation was used on polyimide films
US 4,634,755 (American) 1987	Method for making norbornane anhydride substituted polyorganosiloxane	General Electric; JF Hallgren, DV Brezniak	Siloxane-imide block copolymer useful as elastomeric adhesives or as epoxy curing agents	A method is provided for making norbornane anhydride substituted organosiloxane by equilibrating norbornane anhydride organosiloxane with cycloorganosiloxanes in the presence of an acidic catalyst

Table 2. (continue)

Patent No.	Title	Assignee; Inventor	Uses/Advantages	Technical Summary
US 4,670,497 (American) 1987	Soluble silicone-imide copolymers	General Electric; CJ Lee	Protective coatings for semiconductors and electronic devices; processable over a wide temperature range	Diglyme soluble siloxane-imide copolymers with blocks of widely differing T_gs. Examples based on BTDA, PMDA, BPADA, SDA, TDA, MDA, ODA, GAPDS, GAPPS-9
US 4,558,110 (American) 1985	Crystalline silicone-imide copolymers	General Electric; CJ Lee	Copolymers useful as protective coatings for semiconductors and electronic devices; have lower melt viscosity	Crystalline siloxane-imide block copolymers were prepared from aromatic dianhydrides and bis-amino-terminated siloxanes.
JP 59004623 (Japanese) 1984	Poly(amic acid)-silicone intermediates	Hitachi Chemical Co. Ltd., Japan	Coatings showed excellent adhesion to glass	Polysiloxane-imides were prepared from BPDA, aromatic diamines and aminopropylsiloxane and heat imidized to form heat-resistant coatings with excellent adhesion
US 4,011,279 (American) 1977	Process for making polyimide-polydiorganosiloxane block copolymers	General Electric; A Berger, PC Juliano	Block copolymer useful as an insulating layer on semiconductor devices	A dianhydride and diamine in the presence of an acid catalyst produces an intermediate polyimide product; an amino-terminated siloxane is added at a reduced temperature
FR 2236887 (French) 1975	Poly(ether imides)	General Electric; DR Heath, JG Wirth	Moldable, heat-stable polyimide-siloxanes	
US 3,887,636 (American) 1975	Organo(block-amide-siloxane)-(block-amide-imide) polymers	General Electric: PC Juliano, TD Mitchell, SW Kantor	Useful as coatings on glass and metal conductors for high-temperature, weathering and corona; esp. on wire and cable	Copolymers are prepared by the reaction of a carboxy end-blocked diorganopolysiloxane, and organodicarboxy-terminated imide and an organodiisocyanate to yield high MW linear polymers
US 3,740,305 (American) 1973	Composite material bonded with siloxane-containing polyimides	General Electric: JT Hoback, FF Holub	Copolymers useful to obtain good adhesion to organic thermoplastic, metals and ceramics	Composites were prepared in situ with siloxane/amic acid block copolymer

Table 3. Weight change incurred by homopolymers, poly(siloxane imide) segmented copolymers, and blends with polybenzimidazole due to exposure to high energy, high flux atomic oxygen[a]

Polymer system	Weight loss × 10^4 (g)
Polybenzimidazole	23.13
Homopolyimide	14.01
Poly(siloxane imide)-segmented copolymer 10% siloxane	4.54
Poly(siloxane imide)-segmented copolymer 40% siloxane	+3.10
Blend of poly(siloxane imide) copolymer (75%) and PBI (25%); total siloxane content 30%	+2.95
Blend of poly(siloxane imide) copolymer (60%) and PBI (40%); total siloxane content 24%	+2.30
Kapton	22.93
Kapton coated with ~1500 Å poly(siloxane imide) copolymer with 50% siloxane	5.96

[a] All polyimides based on BTDA and *m*DDS; all poly(siloxane imide)-segmented copolymers have siloxane segment molecular weight of 950 g/mol; all compositions expressed on a weight basis

type measurements, very high concentrations of siloxane, e.g. nearly 90%, are often observed, even though there may be as little as 5 or 10% of the siloxane actually in the bulk copolymer. The thermodynamic driving force was discussed by Dwight as well as in other papers [75,119]. The application of this phenomena then depends upon the siloxane being near the surface, where high temperatures in air or aggressive oxygen transform the organosiloxane to at least a partially, organosilicate structure. Thus, the phenomena that occurs with atomic oxygen may be formally similar to burning in air described earlier. This was also demonstrated in additional papers produced by Arnold and co-workers. The systems utilized were either pure imide-siloxane copolymers or miscible blends of certain polyimides with other engineering polymers such as polybenzimidazoles [120–122] (see Fig. 20). Here the fact that the two engineering polymers are mutually miscible provided a mechanism for "anchoring" the siloxane phase, which nevertheless still tries to drive to the air or vacuum interface. In fact, weight loss measurements in atomic oxygen showed that the siloxane could also protect these engineering thermoplastic blend systems from aggressive oxygen, including oxygen plasmas (see Table 3). Indeed, some of the oxidized copolymer blend systems showed a net weight gain! In a different but related study, Young et al. [123] reviewed a NASA program called LDEF, which investigated the role of outer space on the degradation of materials, during a relatively long 69-month period. One of the conclusions from this study was that the silicone-modified imide systems did provide a significant degree of protection relative to other organic materials.

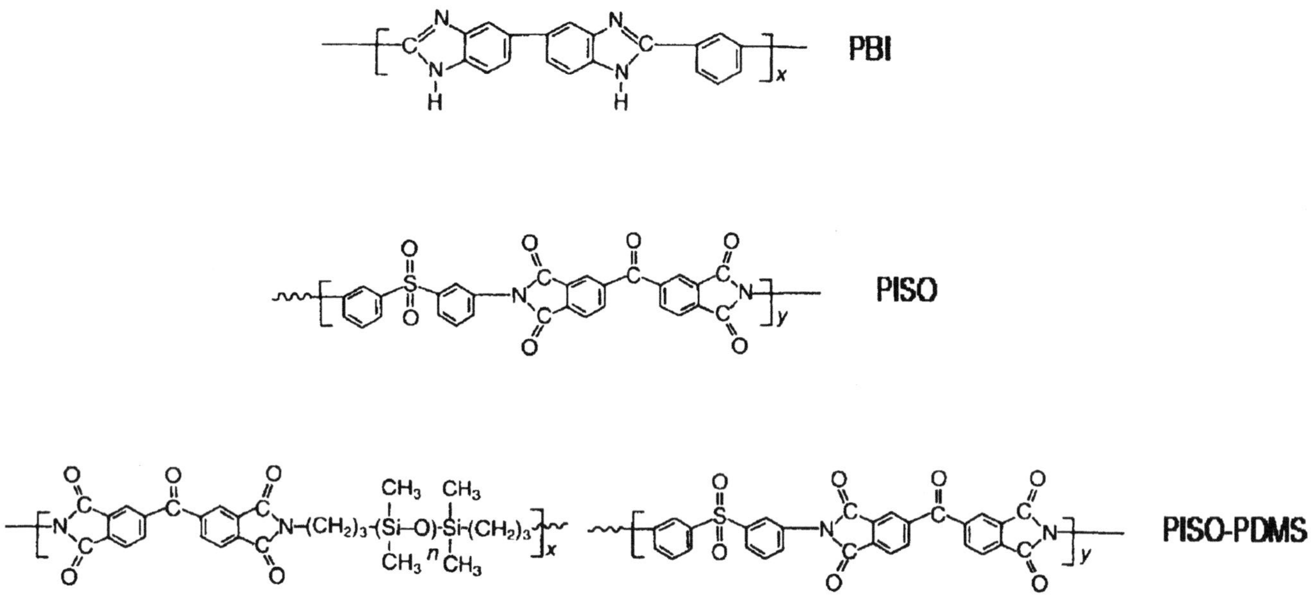

Fig. 20. Polymers used for blending studies

Table 4. Poly(aryl imide) homo- and copolymers which resulted in miscible blends with PEEK

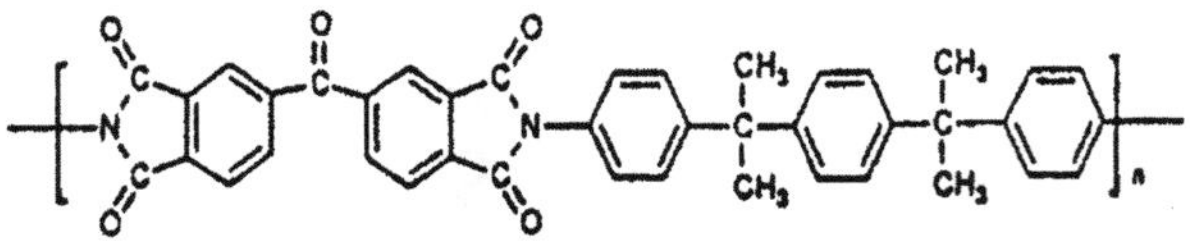

BTDA - Bis P based Polyimide

Homopolymer

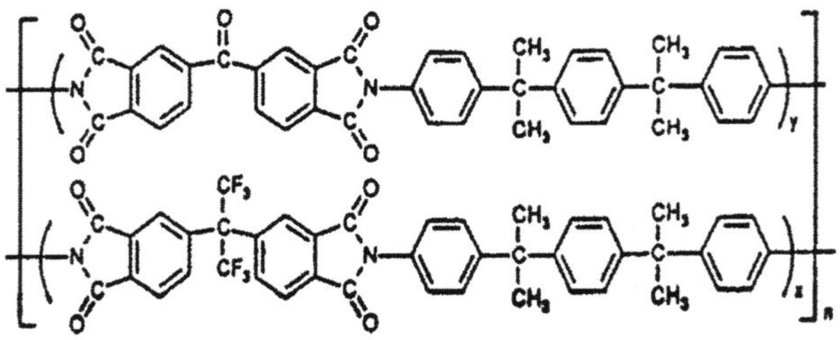

BTDA/6FDA - BIS P based Polyimide

Random Copolyimide

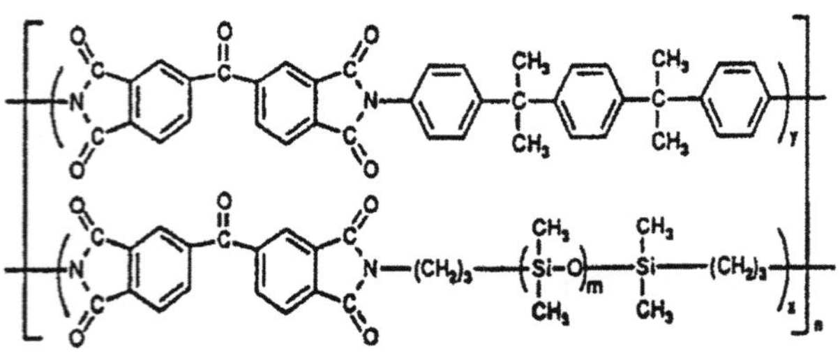

BTDA - Bis P/PSX based Polyimide

Segmented Polyimide-PDMS Copolymer

In addition to the polybenzimidazole blends with imide-siloxanes, it was also demonstrated that other different polyether imide-siloxane-segmented copolymers of variable composition could be used to modify semi-crystalline polyether ketones (PEEK) by presumably an analogous mechanism. The mechanical properties and transition temperatures of those systems have been reported [124,125], and would appear to be additional suitable candidates for the harsh outer space environment (see Table 4 and Fig. 21).

Some new electronic applications of polyimide-siloxane adhesives have been already partially reviewed. In addition, the development of photo-crosslinkable copolymers has been of great interest for the development of components of electronic devices [126]. The general field of photosensitive polymers was reviewed by Horie and Hamishita in 1995 [127]. A number of papers have investi-

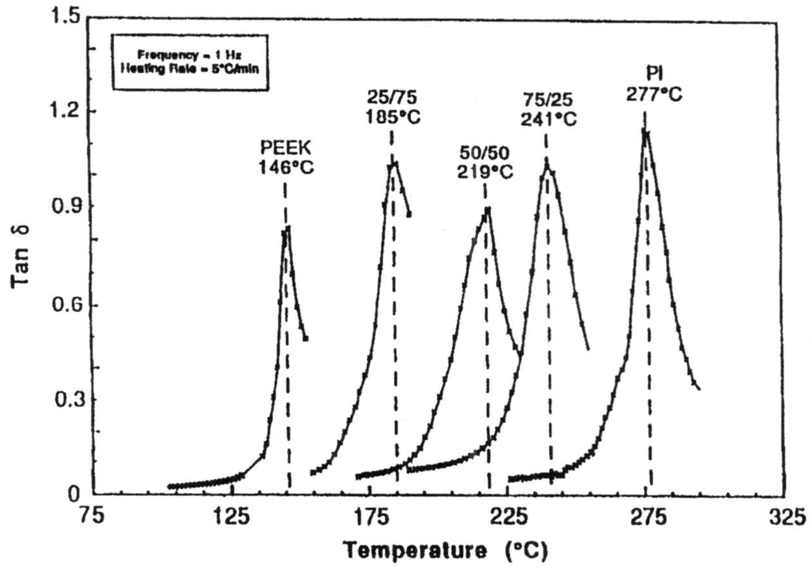

Fig. 21. Dynamic mechanical behavior of miscible, amorphous PEEK/polyimide (BTDA-Bis P) blends (composition in wt%)

gated the development of photo-crosslinkable polyimides which contain silxoane segments. Moyer et al. [128] developed a series of soluble high molecular weight photo-imagable copolymers which exhibited high glass transition temperatures, good photosensitivity properties, and satisfactory thermal stability. The photosensitivity was based on variations of known techniques, i.e. the introduction of benzylic dimethyl-substituted aromatic amines as a component of the imide, which in the presence of a ketone component, such as BTDA, can be excited by UV light. The radicals generated can recombine to produce a crosslink or "negative" device component. The optical densities of the siloxane-modified polyimides were desirably very significantly decreased with increasing siloxane content, without a significant decrease in the photosensitivity of the resulting system. This improved optical clarity allowed thicker films to be crosslinked at lower exposure doses. Additional work by Moyer [129] involved utilization of the "6F dianhydride" which was also able to enhance optical clarity in desirable ways. The synthesis and characterization of photosensitive silicone-containing copolyimides were also investigated by Li et al. [130].

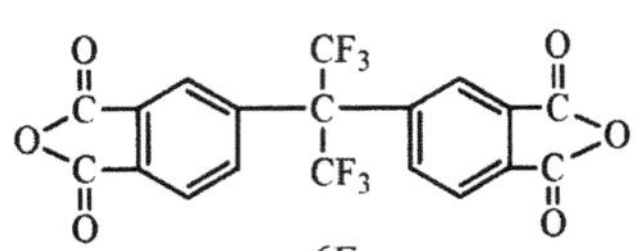

6F

A variety of other efforts have been reported, including interesting work by Tunney and Fitzgerald [131], on the utilization of fluorinated imide-siloxane materials. These may also have applications for thermal dye transfer [132]. The photo-crosslinking reactions of benzophenone-containing polyimide-siloxanes have also been investigated in solution by Shindo et al. [133]. Solution reactions would partially favor cyclization vs. crosslinking, as expected. Photosensitive polyimide-siloxanes have also been developed into compositions that have been reported to be useful as insulating films by Ishikawa et al. [134]. An important paper by Volksen, Hedrick, Russell and Swanson [135] illustrates the design and synthesis of an imide-siloxane block copolymer in their utilization as an oxygen ion etch barrier. Again, it is known from earlier studies that the aggressive oxygen can transform the siloxane into silicate which in this case is very useful as an ion etch barrier. Further modifications of the copolymers have been investigated by a number of workers. For example, Saraf et al. [136] have reported the development of electrically conducting compositions. This is somewhat related to additional efforts in the area of conductive adhesives [137]. For example, Nguyen and Wood [138] reported the utilization of silver-filled copolymers as die attach materials. The longevity or wear of the copolymer coatings is of great interest and this has been studied by Eiss and Kang [139] who investigated a particular type of wear known as fretting wear of these copolymers in coating form. They demonstrated the utility of the siloxane in coatings as a function of varying humidity. This work was consistent with earlier studies by Kaltenecker-Commercon [140] who reported a fundamental study of water aggression in polyimide-siloxanes, which was also published separately, as indicated earlier. The modification of the copolymers for electrical purposes is certainly of continuing interest and Tian, Pak and Xu reported that they could demonstrate high temperature polyelectrolyte-type behavior from the imide-siloxane copolymers [141]. However, despite the utilization of conducting formulations, very significant microelectronics interest remains in developing polyimide copolymers that will show improved low dielectric behavior. In this connection, Arnold et al. [142] demonstrated that the siloxane modification of even fluorinated imides could provide slightly lower values than the pure homopolymers. Variations on this approach will no doubt continue to be of interest.

4
Conclusions

Segmented polyimide-polydimethylsiloxane copolymers have been successfully synthesized both in laboratory and industrial quantities to produce multiphase siloxane-modified polyimides. The siloxane detracts somewhat from the otherwise excellent thermooxidative stability of the polyimide, but it does produce a number of important properties. These include multiphase behavior, improved adhesion to many substrates, improvements in fire resistance and enhanced gas and liquid separation membranes, where one wishes not only to maximize the contribution of the siloxane to permeability, but also to utilize the imide to retain some selectivity and rigidity in the thin film membranes. Aerospace and electronic applications for the copolymers and blends have also been developed which are partially dependent on the hydrophobic nature of the siloxanes and that they can be converted into partially inorganic-like silicate surfaces, either under very high temperatures in air, or through the initiation of reactions with aggressive oxygen, such as oxygen plasmas or atomic oxygen. The resulting products may have important applications as ion etch resistant materials for electronics and for environmentally stable atomic oxygen resistant materials for spacecraft and possibly satellite systems. The influence of structure on the stability of the materials has been examined and aromatic amine functional systems are at best only moderately better than the traditionally used aminopropyl functional oligomers. Further developments can be expected in newer areas such as inorganic-organic hybrids, new adhesives, encapsulants and membranes.

Acknowledgments. James E. McGrath would like to thank several funding organizations for support of our part of this research over a considerable period of time. These include the National Science Foundation, the Office of Naval Research, the Air Force Office of Scientific Research, NASA-Langley, the Army Research Office, the IBM Corporation and the GenCorp Foundation. We would also like to thank all of the colleagues and especially current and former students who have contributed to this body of knowledge.

5
References

1. Mittal KL (ed) (1984) Polyimides: synthesis, characterization and applications, vols 1 and 2. Plenum, New York; Ghosh MK, Mittal KL (eds) (1996) Polyimides: fundamentals and applications. Marcel Dekker
2. Feger C, Khojasteh MM, McGrath JE (eds) (1989) Polyimides: materials chemistry and characterization. Elsevier, Amsterdam
3. Wilson D, Stenzenberger HD, Hergenrother PM (eds) (1990) Polyimides. Blackie, Glasgow
4. Lupinski JH, Moore RS (eds) (1989) Polymeric materials for electronics packaging and interconnections, ACS Symposium Series 407. American Chemical Society, Washington DC
5. Takekoshi T (1990) Adv Poly Sci 94 1
6. Cassidy PC, Fawcett NC (1982) In: Grayson ME (ed) Encyclopedia of chemical technology. Wiley, New York, 18: 704
7. Verbicky JW Jr. (1988) In: Mark HF, Bikales NM, Overberger CG, Menges G (eds) Encyclopedia of polymer science and engineering, 2nd edn. Wiley, New York 12: pp 364–383
8. Sato M Polyimides (1997) Plast Eng (NY) 41 (Handbook of Thermoplastics) 665–699,Marcel Dekker
9. Yilgor I, McGrath JE (1988) Polysiloxane containing copolymers: a survey of recent developments. Adv Polym Sci 86:1
10. Noshay A, McGrath JE (1977) Block copolymers: overview and critical survey. Academic Press, New York
11. Greber G (1971) J. Praktische Chem Band 313:3 461
12. Greber G (1968) Angewante Makro Chemie 4/5:41 212–254
13. Kuckertz VH (1966) Makromol Chem 98:101

14. Hoback JT, Holub F (1971) Siloxane-containing prepolymers for making poly(amide imides). US Patent 3723385 5 pp
15. Hoback JT, Holub FF (1971) Composite materials bonded with siloxane-containing polyimides. US Patent 3740305 5 pp
16. Maudgal S, St Clair TL (1984) Preparation and characterization of siloxane-containing thermoplastic polyimides. Int J Adhes 4:2 87
17. Maudgal S, STClair TL (1984) Siloxane containing addition polyimides ii acetylene-terminated polyimides SAMPE Quarterly 16:1 6–12
18. Yilgor I, Yilgor E, Johnson BJ, Eberle J, Wilkes GL, McGrath JE (1983) Segmented polysiloxane-polyimide copolymers. Polymer Preprints 24(2):78
19. Yilgor I, Yilgor E, Eberle J, Steckle W Jr, Johnson BC, Tyagi D, Wilkes GL, McGrath JE (1983) Novel segmented elastomers from amino alkyl-terminated dimethylsiloxane oligomers Polymer Preprints 24(1):167
20. Johnson BC (1984) PhD thesis, Materials Engineering Science. High-performance polyimide copolymers: synthesis and characteristics. Virginia Polytechnic Institute and State University
21. Johnson BC, Yilgor I, McGrath JE (1984) Synthesis of poly(imide)-polysiloxane segmented copolymers. Polymer Preprints 25(2):54
22. McGrath JE (1985) In: McGrath JE (ed) Ring opening polymerization: kinetics mechanisms and synthesis, ACS Symposium Series No 286. American Chemical Society, Washington DC, pp 1–22
23. General Electric Company (1967) US Patent 3,325,450
24. Eddy VJ, Hallgren JE (1987) J Org Chem 52:1903
25. Davis GC, Heath BA, Gildenblatt G (1984) Polyimide-siloxane: properties and characterization for thin film. Electronic Applications in Ref 1:847–869
26. Policastro PP, Lupinski JH, Hernandez PK (1988) Siloxane polyimides for interlayer dielectric applications, Polymeric materials: Science Engineering 59:209
27. Bolon DA, Hallgren JE, Eddy VJ, Codella PJ, Davis GC, Regh KA (1989) Patternable silicone polyimide copolymers. In: Feger C, Khojasteh MM, McGrath JE (eds) Polyimides: materials chemistry and characterization. Elsevier, Amsterdam, p 103
28. Lupinski JH, Policastro PP (1990) Polysiloxaneimides. Polym News 15(30):71
29. Cella JA, Gallagher PE, Shank GK Block silicone-polyimides and methods for preparation. Eur Pat Appl EP 295561 A2 881221 13 pp
30. General Electric Company (1983) US Patent 4,381,396
31. Saam JC, Spier JL (1959) J Org Chem 24:119
32. Gilbert AK, Kantor SW (1959) J Polym Sci 40:35
33. Noll W (1968) Chemistry and technology of silicones. Academic Press, New York
34. Rochow EG (1987) Silicon and silicones. Springer, Berlin Heidelberg New York
35. Sormani PM, McGrath JE (1985) Kinetics and mechanisms of the anionic ring opening polymerization of cyclosiloxanes in the presence of bis(1,3-aminopropyl tetramethyldisiloxane) In: McGrath JE (ed) Ring opening polymerization: kinetics mechanisms and synthesis. ACS Symposium Series No 286
36. McGrath JE, Sormani PM, Elsbernd CS, Kilic S (1986) Makromol Chem Macromol Symp 6:67
37. Elsbernd CS, Spinu M, Krukonis VJ, Gallagher PM, Mohanty D K, McGrath JE (1990) Synthesis and fractionation studies of functionalized organosiloxanes In: Ziegler JM, Gordon FW Fearon (eds) Silicon-based polymer science: a comprehensive resource. ACS Advances in Chemistry Series No 224, pp 145–164
38. Speier JL, Roth CA, Ryan, RW (1971) J Org Chem 36 3120; Burns GT, Decker GT, Roy AK (1994) US Patent 5 290 901 to Dow Corning
39. Berger A (1985) Modified polyimides by silicone incorporation Natl SAMPE Symp Exhib (Proc) 30 (Adv Technol Mater Processes) pp 64–73
40. Babu JR, Sinai-Zingde G, Riffle JS (1993) J Poly Sci Chemistry ed 31, 7:1645

41. Sysel P, Oupicky D (1996) Polyimide-polysiloxane block copolymers synthesized from alpha omega-(3-aminophenoxy)-terminated poly[oxy(dimethylsilyl)-14-phenylene (dimethylsilylene)]s Polym Int 49(4):275
42. Sysel P, Babu JR, Konas M, Riffle JS, McGrath JE (1992) Preparation and evaluation of arylamine containing polysiloxane-polyimide block copolymers Polym Prepr (Am Chem Soc Div Polym Chem) 33:2:218
43. Keohan FL, Hallgren JE (1990) Novel poly(imide-siloxane) polymers and copolymers. Adv Chem Ser 224 (Silicon-Based Polym Sci) 165
44. Smith S D (1991) MS thesis. Synthesis and characterization of perfectly alternating segmented copolymers comprised of poly(dimethylsiloxane)s and engineering thermoplastics. Virginia Polytechnic Institute and State University, Blacksburg
45. Summers JD, Elsbernd CS, Sormani PM, Brandt PJA, Arnold CA, Yilgor I, Riffle JS, Kilic S, McGrath JE (1988) Recent advances in organosiloxane copolymers. In: M Zeldin, KJ Wynne, HR Allcock (eds) Inorganic and organometallic polymers: macromolecules containing silicon phosphorus and other inorganic elements, ACS Symposium Series 360, pp 180–198
46. Arnold CA, Summers JD, Chen YP, Bott R, Chen D, McGrath JE (1989) Structure-property behavior of soluble polyimide-dimethylsiloxane segmented copolymers. Polymer 30(6):986
47. Arnold CA, Summers JD, Bott RH, Taylor L T, Ward TC, McGrath JE (1987) Structure property behavior of polyimide-siloxane segmented copolymers. SAMPE Proceedings 32:586
48. Arnold CA, Summers JD, Bott RH, Taylor L T, Ward TC, McGrath JE (1987) Structure-property relationship of polyimide-siloxane copolymers prepared by bulk and solution imidization techniques. Polym Preprints, ACS National Meeting 28(2):217
49. Arnold CA, Summers JD, Chen Y P, Chen DH, Graybeal JD, McGrath JE (1988) Structure-property behavior of soluble polyimide-polydimethylsiloxane segmented copolymers. SAMPE Proc 33:960
50. Arnold CA Summers JD, McGrath JE (1989) Synthesis and physical behavior of siloxane modified polyimides. Polym Eng Sci 29(20):1413
51. Kim YJ, Glass TE, Lyle GD, McGrath JE (1993) Kinetic and mechanistic investigations of the formation of polyimides under homogeneous conditions. Macromolecules 26:1344
52. Rogers ME, Glass TE, Mecham S J, Rodrigues D, Wilkes GL, McGrath JE (1994) Perfectly alternating segmented polyimide-polydimethyl siloxane copolymers via transimidization. J Polym Sci Part A Polym Chem 32:2663
53. Hedrick JL, Brown HR, Volksen W, Sanchez M, Plummer CJG, Hilborn JG (1997) Low stress polyimide block copolymers. Polymer 38(3):605
54. Bowens AD, Sensenich CL, Venkatesan V, Robertson MA, McCartney SR, Lesko JJ, Riffle JS (1997) Waterborne polyimides and poly(imide-siloxane)s. Polym Prepr, Am Chem Soc. Div Polym Chem, 38(2):632
55. Furukawa N, Yausa M, Yamada Y, Kimura Y (1998) Synthesis and properties of novel thermosetting polysiloxane-block-polyimides with vinyl functionality. Polymer 39(13):2491
56. Szesztay M, Ghadir M (1993) GPC investigation of polyimide-siloxane copolymers. Angew Makromol Chem 209:111
57. Konas M, Moy TM, Rogers ME, Shultz AR Ward TC, McGrath JE (1995) Molecular weight characterization of soluble high performance polyimides. 1. polymer-solvent-stationary phase interactions in size exclusion chromatography, J. Polym Sci, Part A: Polym Phys 33:1429
58. Konas M, Moy TM, Rogers ME, Shultz AR, Ward TC, McGrath JE (1995) Molecular weight characterization of soluble high performance polyimides. 2. validity of universal sec calibration and absolute molecular weight calculation, J. Polym Sci, Part A: Polym Phys 33:1441

59. York GA, Waldbauer RO, Arnold CA, Rogers ME, Gungor A, Rodrigues D, Wilkes GL, McGrath JE (1990) Investigation of the morphological structure and physical behavior of high T_g polyimide thermoplastic homopolymers and siloxane segmented copolymers. 35th International SAMPE Symposium pp 579
60. Saraf RF, Feger C, Cohen YC Structure of poly(imide siloxane) Adv. Polym Sci and Tech. Proc. of the 4th Intl Conf on Polyimides. Feger C, Khojasteh MM, MS Htoo MS (eds) pp 433–440, Ellenville New York
61. Furukawa N, Yamada Y, Furukawa M, Yuasa M, Kimura Y (1997) Surface and morphological characterization of polysiloxane-block-polyimides. J Polym Sci Part A: Polym Chem 35(11):2239
62. Yamada Y, Furukawa N (1997) Preparation and characterization of siloxane-imide block copolymers based on 3,3[1]–4,4[1]-benzophenonetetracarboxylic dianhydride. Polym J (Tokyo) 29(11):923
63. Yamada Y, Furukawa N, Furukawa M. Preparation and uses of siloxane polyimide block copolymers. Eur Pat Appl EP 349010 A1 900103 16 pp
64. Bott RH, Summers JD, Arnold CA, Blankenship CP Jr, Taylor L T, Ward TC, McGrath JE (1988) Poly(imide siloxane) segmented copolymer structural adhesives prepared by bulk and solution thermal imidization. SAMPE J 24(4):7
65. Bott RH, Summers JD, Arnold CA, Blankenship CP Jr, Taylor LT, Ward TC, McGrath JE (1988) Poly(imide siloxane) segmented copolymer structural adhesives prepared by bulk and solution thermal imidization. Int SAMPE Symp Exhibit 33 (Mater–Pathway Future pp 1177–1187
66. Bott RH, Summers JD, Arnold CA, Taylor LT, Ward TC, McGrath JE (1987) Synthesis and characteristics of novel poly(imide siloxane) segmented copolymers. J Adhes 23(2):67
67. Yoon TH, McGrath JE (1992) Enhanced adhesive performance of thermoplastic poly(imide siloxane) segmented copolymer with peek-graphite composites by gas plasma treatment. High Perform Polym 4(4):203
68. Yoon TH, Arnold-McKenna CA, McGrath JE (1992) Adhesion behavior of thermoplastic polyimides and poly(imide siloxane) segmented copolymers: influence of test temperatures J Adhes 39(100):15
69. Yoon TH (1991) Adhesion study of thermoplastic polyimides and Ti-6Ai-4v allow and peek-graphite composites. PhD thesis, Virginia Polytechnic Institute and State University, Blacksburg
70. Yoon TH, McGrath JE (1991) Effect of surface preparation and thermoplastic ahesive structure on the adhesion behavior of peek-graphite composites. Mater Res Soc Symp Proc 190 (Plasma Process Synth Mater) 3: 137–142
71. Yoon TH, McGrath JE (1991) Adhesion study of peek/graphite composites. Int SAMPE Symp Exhib 36(1):428
72. Yoon TH, Arnold CA, McGrath JE (1989) Titanium 6/4 single lap shear adhesive performance of polyimide homopolymers and poly(siloxane imide) segmented copolymers. Mater Res Soc Symp Proc 153 (Interfaces Polym Met Ceram) 211
73. Yoon TH, Arnold CA, McGrath JE (1990) Effect of molecular weight and end-group control on the adhesion behavior of thermoplastic polyimides and poly(imide siloxane) segmented copolymers. 35th International SAMPE Symposium 1892
74. Furukawa N, Yamada Y, Kimura Y (1997) Lap shear bond strength of thermoplastic polyimides and copolyimides. High Perform Polym 9(1):17
75. Zhuang H, Gardella JA Jr, Incavo JA, Rojstacze RS, Rosenfeld JC (1997) Investigation of polyimidesiloxanes for use as adhesives by electron spectroscopy for chemical analysis. J Adhes 63:1, 199
76. Furukawa N, Yuasa M, Omoro F, Yamada Y (1996) Adhesive properties of siloxane-modified polyimides and application for multi-layer printed circuit boards. J Adhes 59:1, 281
77. Kaltenecker-Commercon JM, Ward TC, Gungor A, McGrath JE (1994) Water resistance of poly(imide siloxane) adhesives: diffusion coefficients by gravimetric sorption. J Ad hes 44:1, 85

78. Cole HS, Gorowitz B, Gorczyca T, Wojnarowski R, Lupinski J (1992) Polymeric materials requirements for the GE high-density interconnect process. Mater Res Soc Symp Proc 264 (Electronic Packaging Materials Science VI) 43
79. Kaltenecker-Commercon JM, Ward TC (1993) Water resistance of poly(imide siloxane) adhesives: an IGC surface energetics study. J Adhes 42:1, 113
80. Sakamoto YI, Takeda N, Takeda T, Tokoh A, Tang DY (1992) New type film-adhesives for microelectronics application. Proc–Electron Compon Technol Conf 42nd 2315
81. Sashida N, Hirano T, Tokoh A (1989) Photosensitive polyimides with excellent adhesive property for integrated circuit devices. Proc Electron Compon Conf 39th 167070 82
82. Edelman R (1985) A new high temperature polyimide adhesive use with kapton film. Natl SAMPE Symp Exhib (Proc) 30 (Adv Technol Mater Processes) 230
83. Ying L, Edelman R (1986) A novel thermoplastic polyimide for composite matrix applications. 31st International SAMPE Symposium 1131
84. Rosenfeld JC, Rojstaczer SR, Tyrell JA. Three-layer polyimide-siloxane adhesive tapes PCT Int Appl WO 9421744 A1 940929 24 pp
85. Morishige S, Kaneda K, Terajima K, Takeda T, Sakamoto Y, Suzuki T. Polyimide adhesive compositions for ceramics and semiconductor devices. Eur Pat Appl EP 531867 A1 930317 10 pp
86. Inoue H Takabayash S Muramatsu T, Funagoshi T, Hirano T. Polyimide-siloxane block copolymer-based heat-resistant adhesive compositions. US Patent 5180627 A 930119 11 pp
87. Kanakarajan K, Freuz JA. Flexible multilayer polyimide film laminates and their preparation. Eur Pat Appl EP 474054 A2 920311 21 pp
88. Edelman R, Papanu VD (1989) Thermoplastic film die attach adhesives. US Patent 4994207 A 910219 5 pp
89. Keohan FL, Lewis LN (1985) Heat curable silicone polyimide rubber compositions containing cyclometallized platinum phosphite compounds. US Patent 4634610 A 870106 8 pp
90. Lee CJ Soluble silicone-imide copolymers. (1986) US Patent 4586997 A 860506 9 pp
91. Bauer CL, Farris RJ (1989) Stresses in polyimide films In: Feger C, Khojasteh MM, McGrath JE (eds) Polyimides: materials, chemistry, and characterization. Elsevier, Amsterdam, p 549
92. Furukawa N, Yamada Y, Kimura Y (1996) Preparation and stress relaxation properties of thermoplastic polysiloxane block polyimides. High Perform Polym 8(4):617
93. Nelson GL (1995) Fire and Polymer II ACS Symp No 599
94. Keohan FL, Swint SA, Buese MA (1991) Selective degradation of silicone copolymers and networks. J Polym Sci: Part A: Polym Chem 29:303
95. Wescott James M, Yoon Tae Ho, Rodrigues David, Kiefer Laura A, Wilkes Garth L and McGrath JE (1994) Synthesis and characterization of triphenylphosphine oxide containing poly(aryl imide)-poly(dimethyl siloxane) randomly segmented copolymers JMS.-Pure Appl Chem A31(8):1071
96. Stern SA, Vaidyanathan R, Pratt JR (1990) Structure/permeability relationships of silicon-containing polyimides. J Membr Sci 49(1):1
97. Yamada Y, Furukawa N, Tujita Y (1997) Structure and gas separation properties of silicone-containing polyimides. High Perform Polym 9(2):145
98. Tsujita Y, Yoshimura K, Yoshimizu H, Takizawa A, Kinoshita T, Furukawa M, Yamada Y, Wada K (1993) Structure and gas permeability of siloxane-imide block copolymer membranes 1 effect of siloxane content. Polymer 34(12):2597
99. Yamada Y, Furukawa N, Wada K, Tsujita Y, Takizawa A (1991) Structures and gas separation properties of silicone containing polyimides. Adv. Polym Sci and Tech. Proc. of the 4th Intl Conf on Polyimides. Feger C, Khojasteh, MM and MS Htoo MS (eds) pp 482–449, Ellenville New York
100. Mecham SJ, Sekharipuram VN, Rogers ME, Meyer GW, Kim Y, McGrath JE (1994) Gas permeability of polyimide/poly(dimethylsiloxane) block copolymers. Polym Prepr (Am Chem Soc Div Polym Chem) 35(2):803

101. Mecham SJ, Rogers ME, Kim Y, McGrath JE (1993) Gas permeability of high-performance homo- and copolymers. Polym Prepr (Am Chem Soc Div Polym Chem) 34(2):628
102. Mecham SJ (1994) Gas permeability of polyimide-polydimethylsiloxane block copolymers MS dissertation, Virginia Polytechnic Institute and State University, Blacksburg
103. Li Y, Wang X, Ding M, Xu J (1996) Effects of molecular structure on the permeability and permselectivity of aromatic polyimides. J Appl Polym Sci 61(5):741
104. Kawakami Y, Yu SP, Abe T (1992) Synthesis and gas permeability of aromatic polyamide and polyimide having oligodimethylsiloxane in main chain or in side chain. Polym J (Tokyo) 24(10):1129
105. Chen SH, Lee MH, Lai JY. (1996) Polysiloxane-imide membranes: Gas transport properties. Eur Polym J 32(12):1403
106. Schauer J, Sysel P, Marousek V, Pientka Z, Pokorny J, Bleha M (1996) Pervaporation and gas separation membranes made from polyimide/polydimethylsiloxane block copolymers. J Appl Polym Sci 61(8):1333
107. Kononova SV, Kuznetsov YP, Apostel R, Paul D, Schwarz HH (1996) New polymer multilayer pervaporation membrane. Angew Makromol Chem 237:45
108. Langsam, M (1996) Polyimides for gas separation In: Ghosh MK, Mittal KL (eds) Polyimides: fundamentals and applications. Marcel Dekker, p 697
109. Li Y, Wan X, Ding M, Xu J (1996) Effects of molecular structure on the permeability and perselectivity of aromatic polyimides. J Appl Polym Sci 61(5):741
110. Macheras JT, Bikson B, Nelson JK. Method of preparing membranes from blends of polymers. Eur Patent Application (EP 706819 A2 960417) 9 pp
111. Lai JY, Lee MH, Chen SH, Shyu SS (1994) Poly(siloxane imide) membranes I physical properties. Polym J (Tokyo) 26:12 1360–1367
112. Lai JY, Li SH, Lee KR (1994) Perselectivities of polysiloxane-imide membranes for aqueous ethanol mixtures in pervaporation. J Membr Sci 93(3):273
113. Svetlichnii VM, Denisov VM, Kudryavtsev VV Polotskaya GA, Kuznetsov YP (1991) Soluble ether imide-oligodimethylsiloxane copolymers: kinetic mechanical and gas separation properties of polyimides and other high-temperature polymers. Proc Eur Tech Symp 2nd Abadie MJ, Sillion B (eds). Elsevier, Amsterdam, pp 525–535
114. Tsujita Y, Yoshimura K, Yoshimuzu H, Takizawa A, Kinoshita T, Furukawa M, Yamada Y, Wada K (1993) Structure and gas permeability of siloxane-imide block copolymer membranes 1. Effect of siloxane content. Polymer 34(12):2597
115. Spontak RJ, Samseth J, Bedford SE (1991) Structure in cast films of poly(siloxane-imide) copolymers. Eur Polym J 27(2):109
116. Samseth J, Spontak RJ, Mortensen K (1993) The response of microstructure to processing in a series of poly(siloxaneimide) copolymers. J Polym Sci Part B: Polym Phys 31(4):467
117. Spontak RJ, Williams MC (1988) Microstructural response of siloxane-imide and SBS block copolymers to heat treatment. Polym J (Tokyo) 20(8):649
118. Samseth J, Mortensen K, Burns JL, Spontak RJ (1992) Effect of molecular architecture on microstructural characteristics in some polysiloxaneimide multiblock copolymers. J Appl Polym Sci 44:1245
119. Dwight D, McGrath JE, Lawson G, Patel N, York G (1989) Surface and bulk microphase separation in siloxane-containing block copolymers and their blends: The roles of composition and kinetics from multiphase macromolecular systems. BM Culbertson (ed) Plenum, pp 265–288
120. Arnold CA, Chen DH, Chen YP, Waldbauer RO, Rogers ME, McGrath JE (1990) Miscible blends of poly(siloxane imide) segmented copolymers and polybenzimidazole as potential high performance aerospace materials. High Perfm Polym 2(2):83
121. Chen DH, Chen YP, Arnold CA, Hedrick JC, Graybeal JD, McGrath JE (1989) Compatible blends of segmented polyimide siloxane copolymers with engineering thermoplastics I: polybenzimidazole (PBI). 34th International SAMPE Symposium 1247

122. Arnold CA, Chen D, Chen YP, Graybeal JD, Bott RH, Yoon T, McGrath BE, McGrath JE (1988) Polyimide-poly(dimethylsiloxane) segmented copolymers and polyblends as potential aerospace materials. Polym Mater Sci Eng 59:934
123. Young PR, Slemp WS, Whitley KS, Kalil CR, Siochi EJ, Shen JY, Chang AC (1995) LDEF polymeric materials: A summary of Langley characterization. NASA Conf Publ 3275 (Pt 2 LDEF-69 Months in Space), pp 567–599
124. Hedrick JC, Arnold CA, Zumbrum MA, Ward TC, McGrath JE (1990) Poly(arylene ether ketone)/poly(aryl imide) homo- and polydimethylsiloxane segmented copolymer blends: influence of chemical structure on miscibility and physical property behavior. 35th International SAMPE Symposium, pp 82–96
125. McGrath JE, Rogers ME, Arnold CA, Kim YJ and Hedrick JC (1991) Synthesis and blend behavior of high performance homo- and segmented thermoplastic polyimides Makromol Chem Macromol Symp 51:103
126. Tummola RR, Rymaszewski EJ, Klopfenstein AG (1997) Microelectronics packaging handbook, semiconductor packing, part II, 2nd edn. Chapman and Hall ITP, NY
127. Horie K, Hamashita T (1995) Photosensitive polyimides: fundamentals and applications. Technomic, Lancaster Basel
128. Moyer ES, Mohanty DK, Shaw J and McGrath JE (1989) Synthesis and characterization of soluble photoimageable polyimide and poly(imide siloxane) homo- and copolymers. 3rd International SAMPE Electronics Conference, p 894
129. Moyer ES (1989) Ph.D. thesis, Chemistry, Photo-crosslinkable polyimide and poly(imide siloxane) homo- and copolymers: synthesis and characterization, Virginia Polytechnic Institute and State University, Blacksburg
130. Zhu P, Li Z, Feng W, Wang Q, Wang L (1995) Preparation and characterization of negative photosensitive polysiloxane imide. J Appl Polym Sci 55(7):1111
131. Fitzgerald J, Tunney SE, Landry MR (1993) Polymer 34(9):1823
132. Depalma VA, Sharma R, Tunney SE, Brust DP Slipping layer of polyimide-siloxane copolymer for dye-donor element use in thermal dye transfer. US Patent 5252534 14 pp
133. Shindo Y, Hasegawa M, Sonobe Y, Sugimura T (1994) Photo-crosslinking reaction of polydimethylsiloxane/benzophenone-containing polyimides in solution. J Photopolym Sci Technol 7(2):285
134. Ishikawa S, Yasuno H, Sakurai H Photosensitive polyimidesiloxanes and compositions and insulating films made thereof. Eur Pat Appl EP 96-111577 960718 13 pp
135. Volksen W, Hedrick JL, Russell TP, Swanson S (1997) Imide-dimethylsiloxane block copolymers: design and synthesis of a permanent buried oxygen ion etch barrier. J Appl Polym Sci 66(1):199
136. Saraf RF, Roldan JM, Gaynes MA, Booth RB, Ostrander SP, Cooper EI, Sambucetti CJ. Electrically conductive compositions. Eur Pat Appl EP 805616 A1 9711105 11 pp
137. Amagai M, Sano H, Maeda T, Imura T, Saito T (1997) Development of chip scale packages (CSP) for center pad devices. Proc Electron Compon Technol Conf 47th, pp 343–352
138. Nguyen MN, Wood JH (1990) Silver filled polyimidesiloxane die attach material. Int SAMPE Electron Conf 4 (Electron Mater - Our Future), pp 291–301
139. Kang C, Eiss NS, Jr. (1992) Fretting wear of polysiloxane-polyimide copolymer coatings as a function of varying humidity wear. Wear 158(1–2):29
140. Kaltenecker-Commercon JM (1992) Water ingression into poly(imide siloxane)s. PhD dissertation. Virginia Polytechnic Institute and State University, Blacksburg
141. Tian SB, Pak YS, Xu G (1994) Polyimide-polysiloxane-segmented copolymers as high temperature polymer electrolytes. J Polym Sci: Part B: Polym Physics 32:2019
142. Arnold CA, Summers JD, Chen YP, Yoon TY, McGrath BE, Chen D, McGrath JE (1989) Soluble polyimide homopolymers and poly (siloxane imide) segmented copolymers with improved dielectric behavior polyimides In: Feger C, Khojasteh MM, McGrath JE (eds) Materials chemistry and characterization. Elsevier, Amsterdam

Received: August 1998

Polyimide-Epoxy Composites

K. O. Gaw and M. Kakimoto[1]

Department of Organic Materials, Tokyo Institute of Technology, Tokyo 152, Japan
[1]*E-mail: mkakimot@o.cc.titech.ac.jp*

Both polyimides and epoxy have found widespread use in many diverse industries. Polyimides, due to their excellent thermal and chemical stability and low dielectric constant, have become a favorite of the electronics industry. This broad class of polymers has found widespread use in a variety of electronic applications and the low cost has made epoxy popular as matrix materials in composites, adhesives and coatings.

Both polyimide and epoxy possess a number of unusually valuable characteristics that make them amenable for use in a vast array of applications. Until quite recently, however, the two compounds have not been concurrently fabricated into a material that combines their desirable properties. Recently advances in polyimide solubility and have made the fabrication of polyimide-epoxy composites possible. This chapter reports the various fabrication strategies and problems encountered in synthesis and characterization of the formed composites and opens up the possibility of further endeavor in this field of research.

Keywords. Polyimides, Epoxy, IPN, Polyamic acids, Adhesion

1 General Background to Polyimides-Epoxies

Polyimides are noted for their inertness to a wide variety of solvents and chemical and thermal environments. They typically have glass transitions above 250 °C, making them ideal materials for use in high temperature applications. They are also incredibly tough materials and have been used as coatings and adhesives on a wide variety of substrates ranging from metals to glasses. However, they have the drawback of not being able to be easily fabricated in bulk forms. The imide structure is quite stiff in polymeric terms and does not allow easy plastic deformation to occur. This also limits the solubility of these polymers. Consequently, polyimides are typically formed from their soluble and more easily processable precursor polyamic acids (PAA) with a thermal or chemical imidization step being required before the polyimide is obtained.

During imidization, however, severe shrinkage problems can result from the removal of condensation water and residual solvent. This shrinkage limits the processability of most polyimides to those shapes, such as thin films, with large surface areas and small volumes. Thus, with the exception of a few new molding compound formulations, they can be considered thermosetting polymers. Additionally, the high cost of many of the precursor monomers has restricted more widespread use of polyimides and has kept the market to those that can sustain higher cost/property ratios such as the electronics and aerospace industries.

Epoxy, on the other hand, is inexpensive, and is most highly regarded for its ease of fabrication with very little shrinkage and the ability to use a diversity of additives to obtain the desired properties. The epoxy addition reaction forms a

Table 1. Typical properties of polyimide and epoxy [1]

	Tensile strength (kPa)	Elongation (%)	Coefficient of linear expansion (cm/cm °C)	T_{max} (°C)
PI	96500	8	5	350
Epoxy	51000	5	2.5	135

T_{max}=resistance to continuous heat

highly nondeformable three dimensional structure, releases no reaction by-products and results in shrinkages of less than 2% for unmodified systems. The mixing of the components is also a straightforward matter of stoichiometrically adding hardener, typically containing amine groups, to the epoxy monomer at room temperature with or without added solvent. Cured epoxy resins are typically quite inert to most chemical attack and the crosslinked network does not swell. The strength of the network is also quite high, although the toughness is quite low. The physical properties can be improved through the use of reinforcement additives that enable energy absorbing mechanisms such as cavitation or crack blunting to become activated. The raw material price of epoxy is also typically of the order of industrial chemicals and they are sold by the ton or kilogram.

The combination of these two polymers up until quite recently, however, has not been an entirely successful scientific endeavor. This has been caused by problems of miscibility resulting primarily from solvent removal. However, recent developments in the synthesis of polyimides and epoxies and a better understanding of epoxy and polyimide reaction kinetics has made it a possibility that these two versatile polymers (Table 1) can be combined to form composites having exceptional properties.

The fabrication of the material that forms when linear polyimides are mixed or coupled with epoxy resins to form three dimensional interpenetrating networks (IPN) is wrought with problems. These can be viewed from a polymer science aspect, where chemically modifying the structure of the components will result in their compatibility or from an engineering viewpoint where modifying existing fabrication methods and formulations will result in the desired composite materials. The following is a summary of research of epoxy and polyimide combinations to date.

1.1 Epoxy Background

The use of epoxy resins in industry extends back over fifty years since their introduction commercially and they have found an extremely wide range of applications as diverse as coatings and adhesives for electronics to use as matrix materials in aerospace composites. It is the use of advanced composites that is pressuring the advancement of epoxy science and the improvement of epoxy based materials physical properties. Epoxies are capable of being formulated with many additives to increase their mechanical and electrical properties However, epoxy resins have traditionally been limited by their intrinsic brittleness stemming from their three.dimensional crosslinked network structure. The improvement of toughness or lowering of the high dielectric constant, without sacrificing modulus, or glass transition temperature, while retaining their relative low cost, is a challenge for many industries ranging from aircraft manufacture to the electronics industry.

Basic theories and fundamental reactions of epoxies have been comprehensively categorized and chronicled by Lee and Neville in their classic work [2]. The use of oligomers as curing agents has also been briefly explored in the literature [3, 4]. However, with the exception of a wide range of publications on functional group-terminated-butanitrile rubbers (such as amine or carboxyl terminated ATBN [5–11], CTBN [12–18], etc. [19]), and other polymers with reactive end groups [5, 20], there has been little work done on the curing of epoxies with polymers that contain the reactive groups within their polymer backbone. This is surprising for it has been seen that the structure and stoichiometry of the curing agents have a profound impact on the nature and properties of the epoxy network [21–27].

The physical mixing of epoxy with an added polymeric material is the most common method used to reinforce the resin. Some of these elastomeric rubbers are able to react with the epoxide and become chemically bound to the epoxy phase. The properties of the systems are dependent upon the final morphology and the level of interfacial bonding present which are in turn dependent on the nature of the rubbery component. This physical mixing of the epoxide with polymers has several inherent problems, such as finding mutually compatible solvents, the removal of the solvents and phase separation during cure of the epoxy. As mentioned, however, epoxy systems cured with polymers that have reactive groups along their backbone, such as polyamic acid, have not been extensively examined. Chemical, as opposed to mere physical, mixing seems to be a promising route to epoxy-polyimide composites.

1.2 Polyimide Background

Polyimides have also been the subject of many review compilations of papers and continue to be the focus of many conferences around the world [28–33]. Having various chemical forms, including isoimides, polyimides have found use in a vast array of applications that in some cases, due to their superior physical characteristics have displaced epoxies, for example, as matrix polymers in aerospace vehicles.

In response to the simultaneous needs for improving epoxy properties and decreasing the cost of polyimide based materials, several approaches have been attempted to combine the two. The physical addition of polyimide to epoxy has been limited by the fundamental immiscibility of polymers containing imide structures. Consequently, the percentage of polyimide that could be incorporated into epoxy has been below 10 wt%. An alternative approach is the modification of the epoxy monomer to contain imide linkages. The efforts of Martinez et al. [34] has led to the successful fabrication of epoxy polyesterimides and the development of the area of opening the oxirane ring with various imide-acids [35–39]. These researchers made quite a thorough investigation into the chemical nature of the systems studied, although little physical characterization data was presented.

The physical properties of systems having imide groups incorporated with epoxy resin was not seen until much later. Researchers in Korea performed microscopic analysis of commercially available polyetherimide (ULTEM 1000) modified epoxy resins [40]. In their work, electron micrographs of the systems showed a two phase system even though their systems were limited to polyimide contents of less then 10 wt%. Compatibilizing polyesterimides with polyetherimides in order to fabricate in situ epoxy composites was addressed by Seo et al. [41].

Recent additions to these efforts comes from Australia, where Morton et al. [42] have developed epoxy hardening agents containing diimide structures for the improvement of environmental resistance and an increase in toughness. All of the researchers who have conducted physical property measurements of the epoxy-polyimide systems concluded that fracture toughness was improved by incorporation of the imide structure. The modification of the epoxy structure by incorporation of imide structures in the epoxy hardener has also been attempted [43]. However, the properties evaluated led the researchers to believe the use of a phosphorylated amine hardening agent improved the properties and not the presence of imide groups. Surprisingly, the use of isoimides for curing epoxy monomers has also not been addressed at the time of writing.

Efforts continue in the Far East, particularly in Japan, by Horie et al., on photosensitive polyimides containing epoxide groups [44, 45]. These studies focus on the chemical amplification of photocrosslinks in the resulting materials for use in making mask materials in silicon chip fabrication.

1.3 Formation of Epoxy-Polyimide Composites

Hay et al. have conducted a wide ranging survey of polyimides, both soluble and insoluble with epoxy, with the goal of forming molecular composites [46]. The insoluble polyimides were incorporated into the liquid epoxy by addition of polyimide powders having a particles sizes of less than 106 micron diameter. In this work, focus on the toughening of the epoxy matrix and the polyimide contents were kept low, below 10 wt% (15 phr) because of processing difficulties. The polyimides had glass transition temperatures of 246 °C or above so any interaction with the epoxy could be discerned via thermal analytical methods such as dynamic mechanical analysis (DMA), and differential canning calorimetry (DSC), although this latter method was found to be less sensitive. They found that in general the fracture toughness of the epoxies were dramatically improved. However, for the polyimides having low T_g this was not the case. The resulting morphologies of the materials also had an influence on the mechanical behavior of the materials. As with Sefton et al. [47] they found the establishment of a good interface between the two components was necessary for improvement of the physical properties. From dynamic mechanical analysis (DMA), they found intermediate T_g values indicating a partial misciblility of the two components. They concluded that this resulted from crosslinks between residual amic acid groups reacting with the epoxy, interchain crosslinking of the epoxy and crosslinks between the epoxy and the polyimide.

2 General Comments on Polymer Blends

In general, many varieties of experimental techniques have been used to fabricate blended polymer systems. Concurrently, a large number of theories have been proposed to explain the morphology development and physical behavior of polymer blends. Consequently, there has been extensive exploration by numerous polymer chemists, physicists, scientists and materials engineers in this area of polymer science [48–54]. Many reviews of a particular type of polymer blend termed "interpenetrating polymer networks" (IPNs) – molecularly interlocked networks of polymers – have been published recently by an abundance of authors including a superb one edited by Paul and Sperling [55]. They report the formation of IPNs of various polymer combinations by the polymerization of two monomers side by side via differing polymerization mechanisms, in addition to IPN formation by the dissolution of one monomer into a polymer with the subsequent initiation of polymerization of the monomer. Both of these methods result in the formation of intermingled polymeric blends.

The aim of developing a new polymer blend is to synergistically combine the properties of the individual polymers resulting in an improved material. A general precondition to this scheme of fabrication, however, is that in order for the final blended material to have the desired properties, the final polymeric phases must form a heterophasic blend, i.e., they must have at least partial thermodynamic immiscibility. This is in contrast to the requirement that the initial reactants must form initially miscible solutions. These conditions do not seem to be met by most polyimide-epoxy systems.

3 Improving Polyimide-Epoxy Compatibility

3.1 Manufacture Methodology

IPNs are fabricated along several general pathways. In one, a linear or slightly crosslinked polymer is swollen by a monomer, followed by polymerization of the monomer. Another method is the utilization of simultaneous graft polymerization reactions resulting in a compatibilized in situ reaction state [56]. Another method that has recently achieved some popularity is the reactive blending of two oligomers. This occurs where two components of an initially homogeneous oligomeric mixture polymerize via different propagation mechanisms to form mutually entangled polymers. Requirements of these systems include the multifunctionality of the compounds and the miscibility of the reactive monomers.

Because the components must initially form miscible solutions or swollen networks a degree of affinity between the reacting components is needed. Therefore, most of the investigations into epoxy IPNs have involved the use of partially miscible components such as thermoplastic urethanes (TPU) with polystyrenes [57], acrylates [58–61] or esters which form loose hydrogen-bound mixtures during fabrication [62–71]. Epoxy has also been modified with polyetherketones [72], polyether sulfones [5] and even polyetherimides [66] to help improve fracture behavior. These systems, due to immiscibility, tend to be polymer blends with distinct macromolecular phase morphologies and not molecularly mixed compounds.

Exclusively mechanically interlocked linear polymer blends, typically, are not thermodynamically phase stable. Given sufficient thermal energy ($T_{use}>T_g$), molecular motion will cause disentanglement of the chains and demixing to occur. To avoid phase separation, crosslinking of one or both components results in the formation of a semi-IPN or full-IPN, respectively. Crosslinking effectively slows or stops polymer molecular diffusion and halts the phase decomposition process.

A molecularly interlocked IPN of epoxy and polyimide was developed by Gaw et al. to form molecular composites of ODA-PMDA polyimide and DGEBA epoxy [73]. In this system the epoxy monomers were homogeneously mixed with a fully polymerized precursor to the polyimide, polyamic acid, that contained reactive groups that could react with the epoxy forming the three dimensional network. This system overcame many of the problems of previous systems by the use of a novel solvent system.

3.2
Improving Polyimide Solubility

The largest impediment to manufacturing molecularly mixed polyimide-epoxy materials has been the limited number of solvents in which polyimide is soluble. The improvement of solubility of polyimides, wherein solubility is increased by the incorporation of large pendant groups containing carboxylic or heterocyclic side groups, has been the focus of much attention, especially the works of Sillion et. al. [74] and others [75–81].

These solubility enhancing side groups are termed cards, with the corresponding polymers called cardo-polyimides [29]. Solubility of the polyimides can also be improved by the use of fluorinated monomers and monomers containing flexible linkages. Sillion and others have used soluble cardo-polyimides in their quest towards incorporating reactive polyimides into epoxy but they achieved only limited success with much difficulty. The main problem encountered was the insolubility of the polyimides with the epoxy requiring the use of surfactants [16]. Low polyimide loadings, usually less than 10% [74] were also dictated due to the vexing removal problem of solvent removal. The polyimide content has thus been limited by the use of high boiling points of the aprotic solvents, typically, dimethylacetamide (DMAc) and *n*-methyl pyrrolidone (NMP), 166 and 202 °C respectively. Consequently, low percentages of polyimide, and thus little accompanying solvent, are used and phase separation, which is often occurs during solvent removal, is noted.

3.3 Development of Volatile Solvent Systems

Problems involved with using typical aprotic solvents such as dimethyl acetamide (DMAc) or *n*-methyl pyrollidone (NMP) (both high boiling point solvents, see Table 2) during polyamic acid synthesis, or more precisely their removal during thermal imidization include void creation in bulk samples, bubbling and irregular surface textures of films, debonding of the PAA/substrate interface and interactions of the solvent with reaction components during cure.

These problems have been largely circumvented by employing a low boiling point solvent system developed by Echigo et al.[82–84] containing tetrahydrofuran (THF) and methanol (MeOH) for the synthesis of polyamic acid. As seen by Gaw et al., THF/MeOH easily solvates both the polyamic acid and epoxy monomers forming stable solutions and is easily removed prior to cure [73]. The THF/MeOH solvent can be completely removed prior to any epoxy curing or imidization steps eliminating the problems associated with DMAc or NMP removal as seen in Fig. 1.

It should be noted that the individual components that react to form PAA are only sparingly soluble in either THF or methanol alone. However, in mixtures ranging from 50/50 to 90/10 THF/MeOH they are both surprisingly soluble and allow high molecular weight PAA to be formed. The intrinsic viscosities of ODA-PMDA polyamic acid (indicative of molecular weight) in this solvent system, can be seen in Fig. 2.

Table 2. The solvents used in PAA synthesis

Solvent	Boiling point (°C)	Dielectric constant	Vapor Pressure at 25 °C
NMP	202	32	4
DMAc	166	37.78	1.3
THF	66.7	5.8	197
MeOH	64.7	32.7	125.03
THF/MeOH (80/20)	65.74	8.95	183

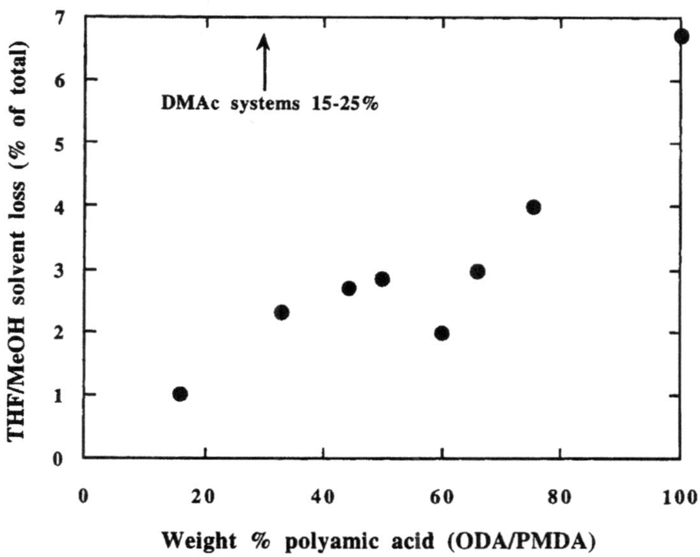

Fig. 1. The loss of solvent vs polyamic acid weight percent in DGEBA/PAA films cast from solution and dried in the atmosphere for 48 h

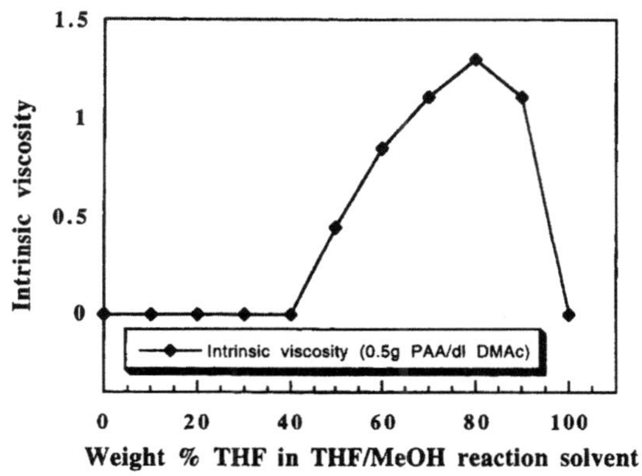

Fig. 2. Intrinsic viscosity of ODA-PMDA polyamic acid as a function of THF/MeOH solvent composition

Using THF/MeOH solvated polyamic acid as a precursor to the highly insoluble polyimide is, to date, the only methodology capable of making epoxy-polyimide composites having polyimide contents greater than 15% [73]. This system shows great promise in manufacturing epoxy-polyimide been fabricated using this solvent system include those derived from monomers listed in Fig. 3 [85].

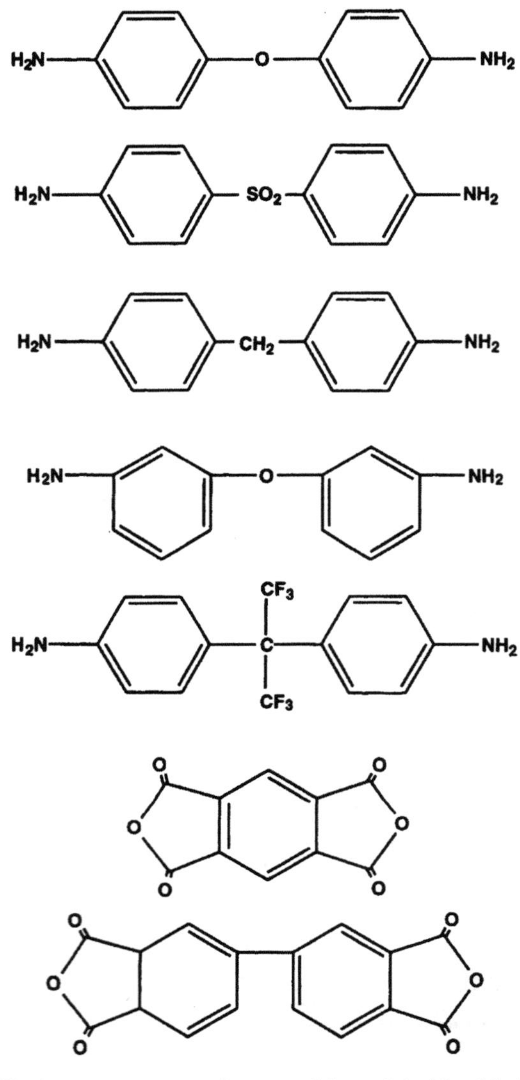

Oxydianiline (ODA) or Diaminodiphenylether (DDE)

Diaminodiphenylsulfone (DDS)

Diaminodiphenylmethane (DDM)

3,3' ODA

6-Fluorodianiline (6FDA)

Pyromellitic Dianhydride (PMDA)

Biphenyl Dianhydride (BPDA)

Fig. 3. Monomer constituents of the polyimides fabricated using 80/20 THF/MeOH as solvent

The solubility of solids in liquids is typically attributed to the dielectric of the solvent, however this does not explain the situation here as is seen in Table 2. The low dielectric constant of the THF/MeOH system should make it a poor solvent for polyamic acid. Consequently, there must be other factors contributing to PAA solubility however, the causes of this solubility behavior have not been elucidated at this time.

3.4 Characterization of Epoxy-Polyimide IPN

Fundamental reaction mechanisms of epoxy cure have been fully investigated in the literature with the work of Lee and Neville being the reference of choice [2]. Recent studies focusing on fundamental reactions of epoxy and various hardeners have been performed using NMR and NIR/FTIR methods in the works of many researchers [86–95]. Mechanisms and kinetics of various epoxy systems has been evaluated and tabulated by Mijovic et al. and others [96, 97].

4 Reactions Between Epoxy-Polyimide Resulting in IPN Formation

As mentioned previously, most of the prior investigations of epoxy systems containing polymeric additives have concentrated on the toughening of the epoxies. A fine review of papers concerning epoxies, toughened, but not cured, with imides is provided by Hay et al. [46]. In all the studies cited therein, the polyimides did not significantly participate in epoxy cure. However, the cure of DGEBA (diglycidyl ether of Bisphenol A) epoxy with ODA-PMDA PAA, as the high molecular weight curing agent, results in an epoxy cured with the amide and carboxylic groups on the backbone of the PAA. Further, upon imidization of the PAA, the polyimide became trapped in the epoxy network and phase separation was be restricted by the existing crosslinked epoxy network.

The simultaneous formation of the polyimide and the three dimensional epoxy network occurs because of the polyamic acid containing reactive groups that can initially open the oxirane groups of the DGEBA and then subsequently intramolecularly cyclize to become polyimide as the epoxy network extends. These reactions are summarized in Fig. 4.

The determination of the exact reaction mechanisms that take place in such a system is hindered by the complexity of the system. Shechter et al. [98] and Dusek and Bleha [99] found that OH generated during reaction of epoxides and amines served as catalytic centers and did not contribute much to cure until there was an excess of epoxide present. These findings were also confirmed by a later report by Prime and Sacher wherein polyamides were reacted with epoxides. However, some generalizations can be inferred from infrared spectra and differential scanning calorimetry studies. The systems tend to form ester linkages first followed by the imidization of the polyamic acid at higher temperatures liberating hydroxyl groups allowing the epoxy matrix to be crosslinked by ether linkages [73, 101].

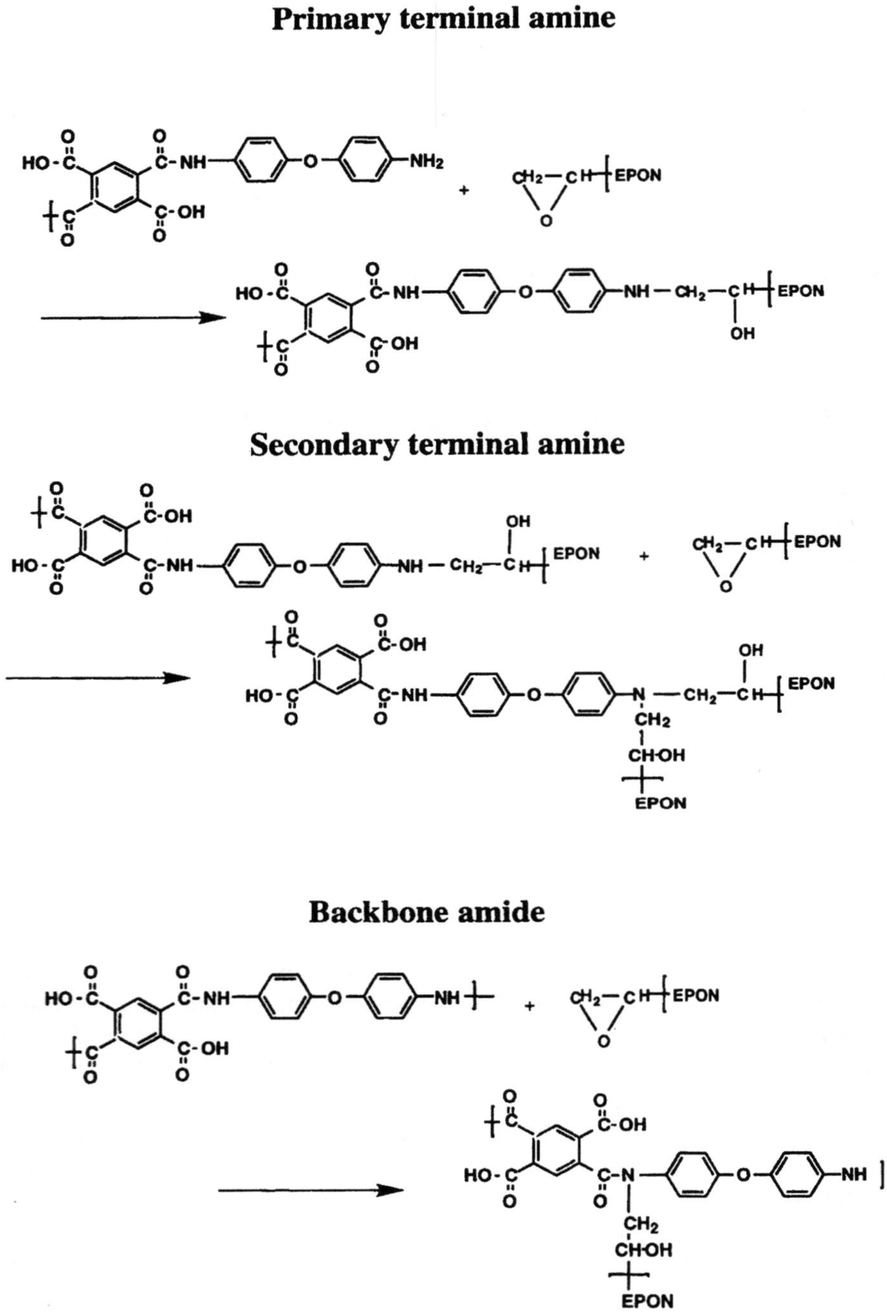

Fig. 4. Possible chemical reactions between epoxy and PAA

Carboxylic acid - ester formation

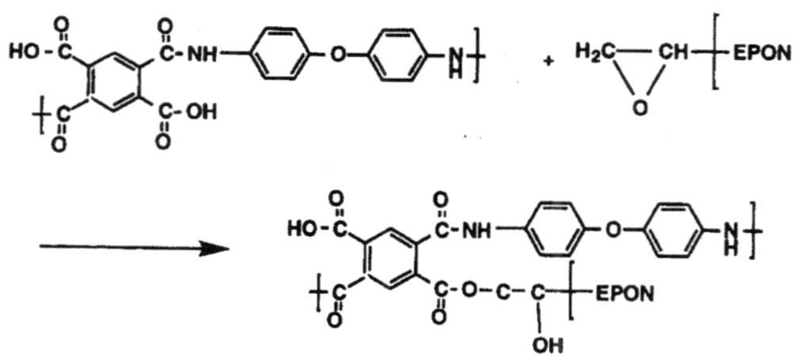

Hydroxyl initiated homopolymerization

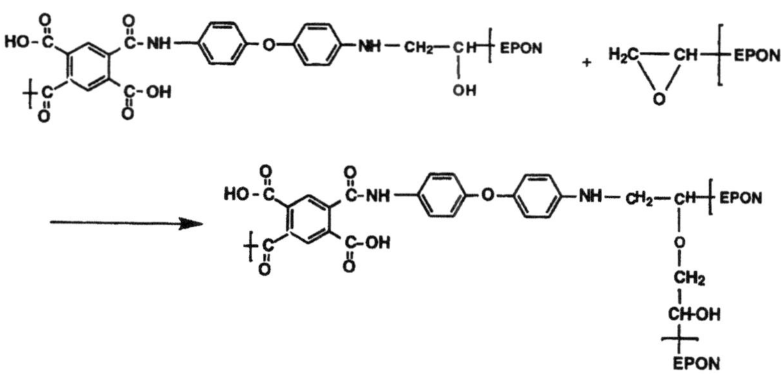

Imidization of ester complex

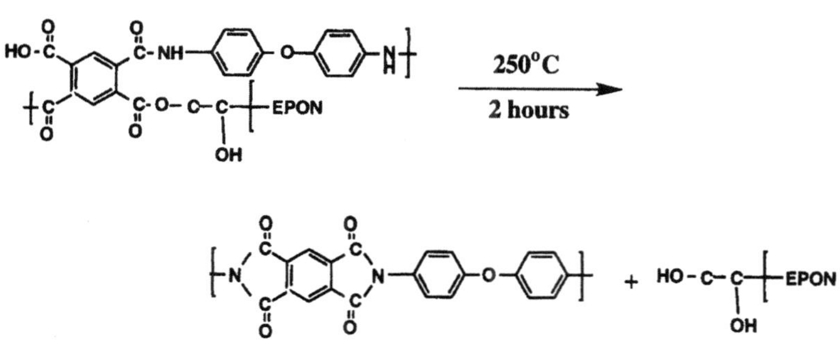

Fig. 4. (continued)

4.1 Morphology of Epoxy-Polyimide Systems

One fine review of the fundamentals of polymer miscibility and the resulting morphology of polymer mixtures is by Coleman et al. and the reader is referred to this for a more complete coverage of the topic [102]. In general, phase separation results from the increase in molecular weight causing a decrease in the configurational entropy of mixing, so that the enthalpy term, which is usually positive (endothermic), becomes more important in determining the free energy of mixing [103]. During the cure of an epoxy network, cured via polyamic acid, the opening of the oxirane ring, along with the transformation of the amic acid groups to imides, will expose more, or less, respectively polar hydroxyl groups, altering the enthalpy of the mixture. Phase separation is controlled by both kinetic and thermodynamic factors. As the reaction proceeds the rates of diffusion will decrease sharply falling off after the gel point has been reached. Consequently, beyond this point a glassy phase exists and phase separation becomes extremely difficult.

5 Glass Transition Temperature

Although the occurrence of fully miscible systems are rare in the literature, as far back as 1965 Krause and Roman reported mixtures of compatible homopolymers exhibiting a single T_g [104]. More recently, Kwei et al. studied polybenzimidazoles and epoxy where full misciblility was seen via microscopic methods [105]. Although limited to methacrylates, Ratzsch gives an adequate treatment of the synthesis and theoretical background of IPNs and reactively coupled polymers focusing on interfacial aspects [106].

Typically, epoxy systems that are mixed with non-reactive polymers and even endgroup reactive polymers phase separate forming diphasic materials [5–20]. This is shown by scanning electron micrographs which revealed large micron sized regions of polymer imbedded in the epoxy matrix. On the other hand, Bucknall et al. have shown that epoxy cured with sulfone containing hardeners and polyethersulfuones (PES), showed mutual compatibility. Scanning electron micrographs of the fracture surfaces of sulfone cured epoxy and PES were featureless, indicating no phase separation. This could be due the common sulfone structure in the two polymers [107–109]. Similar results were found by Maruscelli et al. with super tough epoxy-polycarbonates systems [90].

However, epoxy-polyimide systems employing THF/methanol as the solvent showed that molecular mixing can also be achieved to give featureless micrographs throughout a continuum of polyimide contents ranging from low to high weight contents. However, upon the addition of pyromellitic dianhydride, PMDA (a common epoxy hardener) above stoichiometric amounts, the material showed phase separation. The phase separation was attributed to the faster reaction kinetics of the anhydride with the epoxy compared to those of the polyamic acid [101].

The phase incompatibility of the systems can usually be easily determined by the amount of transparency the systems display. For the above-mentioned systems, totally transparent, albeit colored, composite systems resulted. This showed that at least for these systems phase macroscopic homogeneity was achieved. Typically translucency or cloudiness of films of the materials is indicative of phase separation due to Rayleigh scattering by the domains that have diameters within the range of wavelengths of the visible spectrum.

However, if the refractive indices of the two components are identical Rayleigh scattering will not occur [110]. Due to the various values of refractive indices of epoxy and polyimides quoted in the literature [2] (this is speculated to be a function of crosslink density, monomers used and state of cure) phase separation of epoxy systems can only be inferred and not quantitatively determined visually.

Bucknall, in addition to micrographic studies, confirmed polymer miscibility via dynamic mechanical testing that the T_g peaks in these systems became intermediate to the pure components indicating mutual compatibility [107–109]. This result will be discussed in the following section.

5.1 Glass Transition Temperature Influences

Throughout the history of polymer science there have been efforts to improve (increase) the T_g to increase the useful operating temperature range of polymers. The preponderance of the literature has concentrated on mechanically blended polymeric systems with little component interaction on the molecular level. Where epoxy systems are concerned, the incorporation of additives into the systems results in many changes to the morphology and physical behavior of the material formed.

For a T_g to occur in crosslinked systems, there must be sufficient molecular mobility to affect the macroscopic physical modulus of the material. As the effective molecular weight, between the crosslinks decreases with increasing crosslink density, the thermal activation required to induce sufficient molecular movement, seen by a T_g, is commensurably increased.

The T_g of linear polymers such as polyimides is heavily dependent on, and increases with, the molecular weight. The equivalent statement can also be said about crosslinked networks, where the glass transition temperature is dependent on the network crosslink density [111–117].

The miscibilities of the components in polymer blends is often ascertained by the measurement of the material's glass transition temperature (T_g). The mixing of two polymers with no mutual interactions usually results in the mixture having two separate phases each with their own distinct glass transition temperature. However, when the two components do interact to form a single phase mixture, their glass transitions combine and there will be the emergence of only one transition temperature that is linearly dependent on composition [118].

The discovery of miscible polymer systems, has subsequently inspired the proposal of a number of equations relating the T_g of the individual components to that of the blend [119]. While Gillham focuses on generalized time-temperature-transformation diagrams in his works [120], Karasz and Couchman [121, 122] have proposed the general equation

$$T_g=W_1T_{g1}+W_2T_{g2} \quad (1)$$

where W_x is the weight fraction of component, x. With several assumptions being made this result can be transformed into the well known Fox relation

$$1/T_g=W_1/T_{g1}+W_2/T_{g2} \quad (2)$$

While these equations are crude approximations describing the actual behaviors of various polymeric systems, they have widespread experimental viability [123]. These equations assume that the systems are thermodynamically miscible and stable. Equations resolving the solubility parameters of mixtures can also be done using Small's or Fedors' methods [124, 125]. This latter method only requires the knowledge of the structural formula of the compounds whereas Small's method requires experimental determination of the molar volume.

Although some miscible systems exhibit T_g-composition dependencies as defined by these simple equations, many blends cannot be correlated by them or any of the other well known expressions such as the Kelly-Bueche, Gordon-Taylor or the Gibbs-Dimarzio relationships [126–128]. However, the existence of thermodynamic miscibility has not been proven for epoxy-polyimide systems.

5.2 Component Ratios

The T_g of many epoxy systems has been compiled by Zukas [129]. However, as he found, there are many conflicting reports concerning the effects the stoichiometry of the reactants, typically noted as the amino hydrogen/epoxy ratio or more simply the a/e ratio, on the T_g of the systems. Differences in reaction kinetics are a function of the structures of the epoxy and the hardening agents because these typically determine diffusion and therefore the rates of reaction. Rates of component diffusion decrease during cure due to network formation, and thus the rate and molecular mobility of the epoxy and the curing agent will determine the period during cure in which phase separation can take place, ultimately affecting the crosslink density and the T_g behavior in multicomponent systems.

In general, stoichiometric mixtures of hardening agent to epoxy, amine/epoxy (a/e)=1, gave the highest T_g values [108]. The a/e ratio could not be determined for the epoxy systems that were cured with polyamic acids by the author due to the polydispersity of the amic acids. However, systems containing equal weights of polyamic acid to DGEBA resulted in higher T_g materials.

5.3 Curing Time/Cure Temperature

The curing time profoundly affects the glass transition of epoxy systems; for cure level is dependent on time and temperature of cure. Some systems cured for impractical lengths of time, up to 135 h, exhibited slight increases in the T_g of about 20 °C [108] due to increased crosslink density via etherification.

The T_g of the system is limited ultimately by the curing temperature. One must take care in the use of high temperatures of cure to achieve high T_g and not exceed the degradation temperature of the epoxy or the curing agent. The use of a thermally stable polyamic acid as the curing agent will therefore allow for the use of very high curing temperatures with degradation of the epoxy being the limit of the curing temperature. This is fortunate, for, as Horie and others have noted, as curing temperatures are increased, the T_g of the system increases [130]. However, the crosslinking density reaches a maximum at the cure temperature and therefore the T_g of the system cannot exceed that of its curing temperature.

5.4 Methodology of T_g Determination

T_g values can be determined via either calorimetric, dynamic scanning calorimetry (DSC) or mechanical dynamic mechanical analysis (DMA) measurements. However, since three dimensional highly crosslinked systems have relatively small amounts of molecular motion, the DSC method is not particularly sensitive for T_g determination [131, 132]. Fry and Lind have reported that DSC is misleading, as reactive groups are often sufficiently entrapped in the vitrified structure to give spurious results [133].

In addition, Seferis and Wedgewood have pointed out the many pitfalls that should be avoided when using dynamic mechanical analysis (DMA) to determine thermal properties in epoxy systems [134]. However, Sanz, et al. have investigated T_g of epoxy systems via DMA for a myriad of epoxy compositions and compiled large amounts of reasonable data using this technique [135]. Zukas has done the same using torsional braid analysis (TBA) on many epoxy systems and produced similar conclusions to Sanz [129].

Thus, mechanical measurements such as DMA or TBA are more common with the latter being used on reactive systems to gather reaction kinetics data [120]. These methods relate changes in the responsive modulus of the material to an impressed sinusoidal vibration. From this T_g, the physical thermomechanical behavior of the system can be related by a quantity termed tan δ (storage modulus/loss modulus) which passes through a maximum at the T_g. These relaxations occur at certain frequencies at characteristic temperatures.

5.5 Phase Determination Via Dynamic Mechanical Testing

Examples of DMA of ODA-PMDA polyimide, DGEBA cured with either ODA, can be seen in Figs. 5–7. For comparison, DMA of ODA-PMDA PAA cured DGEBA can be seen in Fig. 9–11. The neat epoxy component exhibited a T_g of 125 or 175 °C (ODA and PMDA cure, respectively) and the polyimide T_g at approxi-

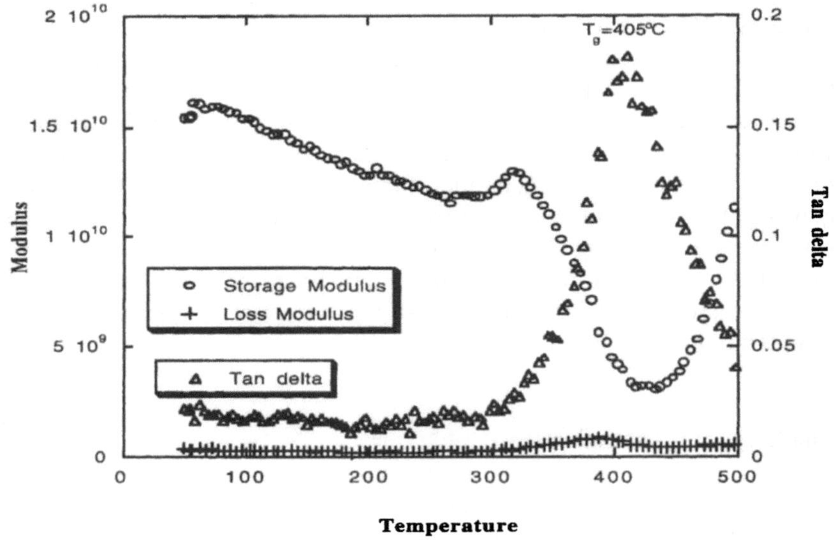

Fig. 5. DMA of ODA-PMDA polyimide

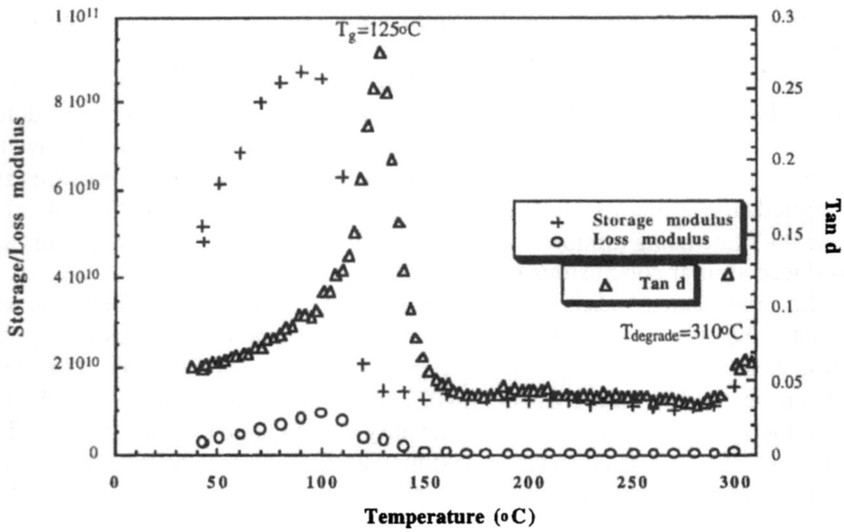

Fig. 6. DGEBA cured with stoichiometric amount (a/e=1)of ODA

mately 405 °C. The interaction of the DGEBA and the PAA can be seen by the migration of the epoxy T_g values to intermediate values ranging from 140–242 °C. Contrary to Latha et al., the T_g values of a crosslinked epoxy resin and the incorporated polyamic acid/polyimide are significantly altered [136]. The level of in-

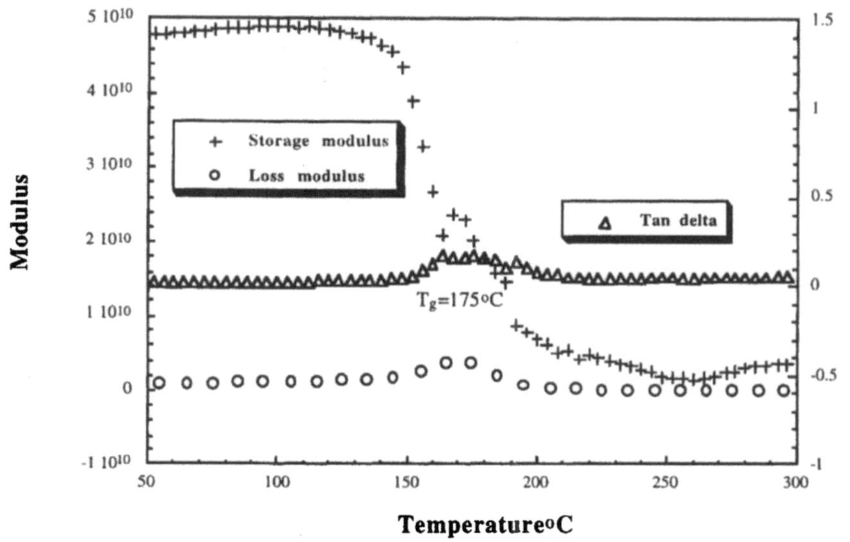

Fig. 7. DGEBA cured with a stoichiometric amount (anhydride/epoxy=1) of PMDA

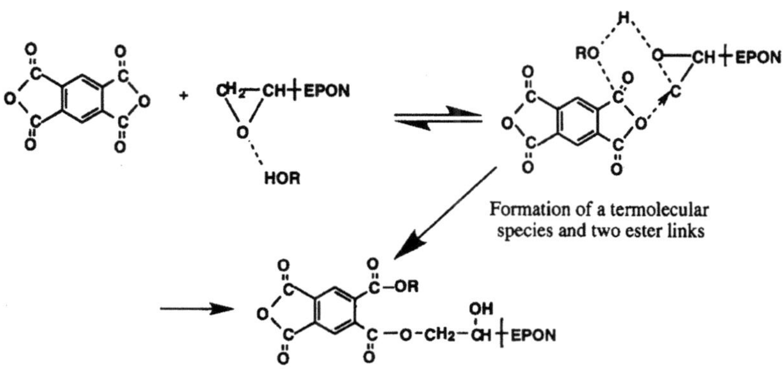

Fig. 8. Diester formation upon cure of epoxy with an anhydride

teraction via reaction between the carboxyl groups, amide groups, etc. of the PAA and the epoxy is significant.

All systems in the figures were cured to apparent completeness as determined by the lack of an exotherm in the DSC of the cured samples [73, 101]. All variations in T_g are the consequence of (I) the extent of molecular interaction resulting from varying the (DGEBA/ODA-PMDA PAA/PMDA) ratios, (II) changes in the network morphology (crosslink density), or (III) a combination of both of these effects.

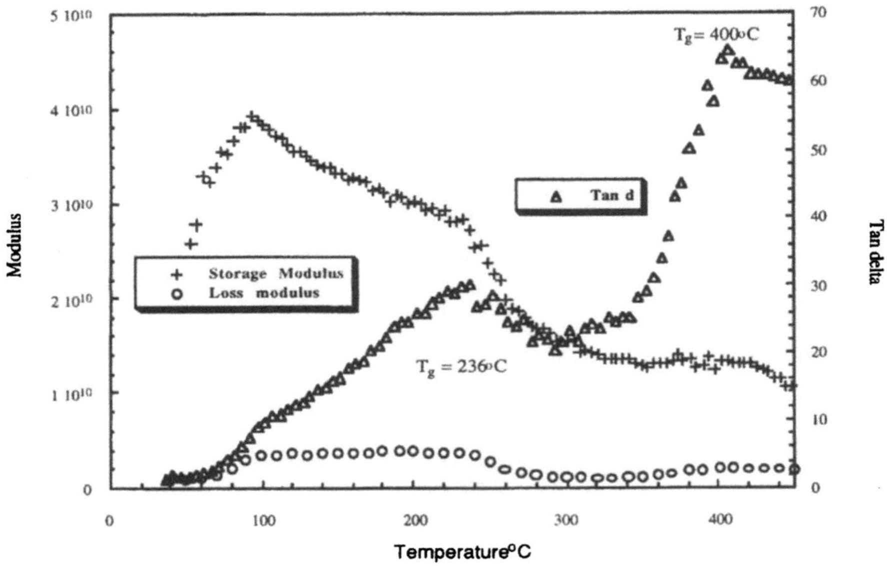

Fig. 9. DGEBA cured with ODA-PMDA polyamic acid (60/40 wt%)

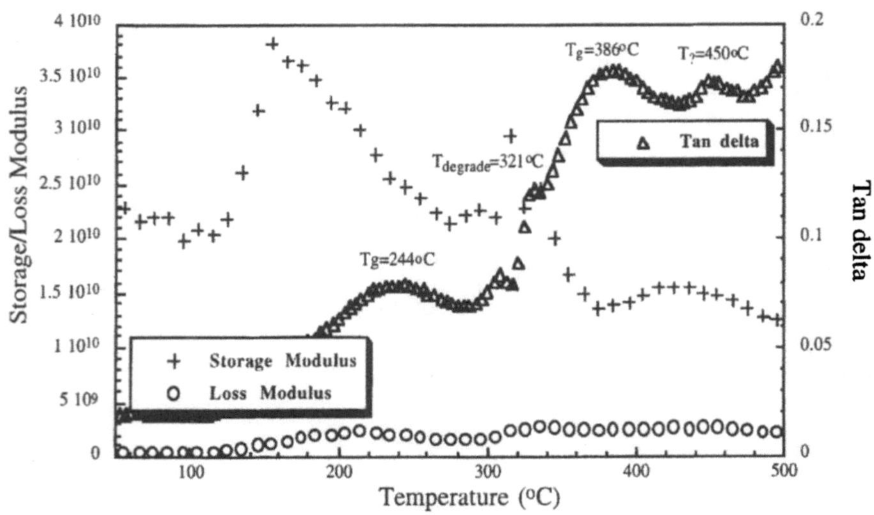

Fig. 10. DGEBA cured with ODA-PMDA polyamic acid and PMDA (22/70/8)

From DMAs of ODA and PMDA cured DGEBA containing no PAA, T_gs of 125 and 175 °C were obtained. These DMA can be seen in Figs. 6 and 7, respectively. The amine was more reactive than the anhydride crosslinks formed extensively at lower temperatures. The PMDA requires the presence of a proton do-

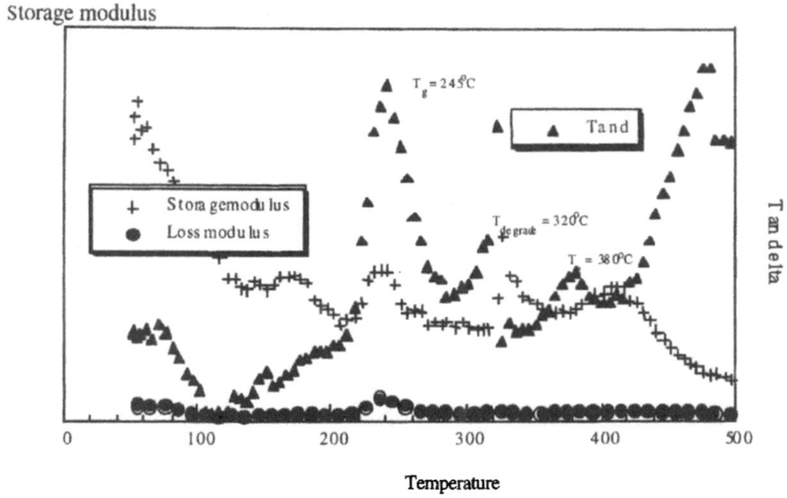

Fig. 11. DGEBA cured with ODA-PMDA polyamic acid and PMDA (38/50/12)

nating group like the carboxyl of the polyamic acid, to initiate cure. When this condition is met the ring opening reaction will proceed via a diester complex as seen in Fig. 8 [136]. As is expected of a lightly crosslinked system, the PMDA-DGEBA systems have broad T_g peaks as compared to the sharp peaks of the ODA cured systems. The T_g of anhydride cured systems is less acute and more oblique indicating a network with a wider distribution of molecular weights between crosslinks.

The DMA of the 60/40 DGEBA/PAA system can be seen in Fig. 9. For all systems, the position of the epoxy T_g peak increases with PAA content indicating increased crosslinking of the epoxy due to increased reactive group concentration. The increase in the epoxy T_g could also indicate a miscibility of the polyimide into the epoxy network.

In the systems cured with both PMDA and PAA, Figs. 10 and 11, the amount of reaction that takes place early during cure has a profound effect on the broadness of the peaks seen. The broadness of the epoxy T_g peak (22/70/8) is due to the increased variability of the epoxy network crosslink density. This phenomena could be caused by the incomplete imidization of the polyimide resulting in a residual polyimide/polyamic ester complex with the epoxy network. This incomplete imidization is due to steric hindrance of the polyamic acid molecule by the surrounding epoxy network not allowing rotation and cyclization of the imide ring. Thus, the polyamic acid is molecularly "locked" into the three dimensional network of the epoxy, becoming an integral part of the network. For the 38/50/12 system, the epoxy T_g reached a maximum at 245 °C, possibly indicating a maximum of polyimide miscibility and of crosslinking density.

In the systems with both PAA and PMDA there were two distinct phases. This is brought about by the kinetics of curing favoring the low molecular weight anhydride to the exclusion of the polyamic acid (Figs. 10 and 11). The lower loss peak increased slightly, from 236 to 245 °C with decreasing DGEBA content (increasing PAA). At high PMDA contents the epoxy reacted in the early stages of cure independently of the kinetically excluded PAA. Consequently, there is very little interaction with the polyimide and molecular vibrations are free of any polyimide influence.

The polyimide T_g value ranged from 380 to 405 °C. The breadth and position of this range is evidence of the inhibition of the imidization of the polyamic acid by the epoxy network. This view is strengthened by the fact that PI T_g increased with decreasing epoxy content. The lack of a T_g at the low PAA content is due to epoxy degradation or the steric hindrance of the polyamic acid molecules by the surrounding epoxy network. The slight decrease in the T_g of the polyimide indicates that the interaction between the epoxy network and the nearly fully imidized polyimide is slight, perhaps indicating a solubility limit. The limited intrinsic solubility of the polyimide in the epoxy, however, produces a very large increase in the epoxy T_g, up to 65 °C higher than either the amine or anhydride cured systems. It is speculated that the combination of anhydride cured systems and the polyamic acid has a synergistic affect on the cure of the epoxy.

5.6 Morphology-T_g Relationship

The relaxations of the system are dependent on the morphologies exhibited by the system as noted in morphology studies of various epoxy systems [2, 136–140]. This was also seen in the studies of DGEBA-PI composites in the authors work with these systems [73, 101].

Phase segregation could account for the presence of the two T_g peaks in the DMA. However, the SEMs of these systems with no PMDA present were featureless [101]. The inability of the polyimide to diffuse to form a separate phase in the crosslinked DGEBA and its ability to segregate from the system when PMDA was added was confirmed via SEM. Thus, it seems that phase segregation takes place during early stages of cure in systems cured with both PAA and PMDA and is limited by the diffusion kinetics of the monomers that react to form the epoxy network and not the diffusion of the polyimide molecules.

From T_g considerations, the polyimide/epoxy system is not miscible when a monomeric hardener is included in the curing of the system. The systems cured exclusively with the polyamic acid were initially single T_g materials at low PAA content, indicating the existence of a single phase. But with increasing PAA content, these systems experienced a phase separation on a molecular scale resulting the evolution of a high temperature relaxation attributed to the polyimide; this morphology, however, could not be determined via microscopic methods.

The two glass transitions became evident due to the severing of the ester bonds formed between the polyamic acid and the epoxy upon the cyclization of the polyimide at higher cure temperatures. The PI molecules then had no direct chemical bond to the surrounding epoxy network. Thus the material forms a diffusionally hindered one-phase material, but on the molecular level two distinct molecular entities exists. This molecular composite of polyimide molecules imbedded in an epoxy matrix resembles a randomly linearly reinforced epoxy network. These results indicate that, with the formation of molecular composites of polyimide and epoxy, one could potentially tailor the T_g by the judicious choice of epoxy and polyimide.

6
Physical Properties of Epoxy-Polyimide Systems

Although several theories attribute the strength of epoxy systems to plastic flow, it is generally believed that epoxy strength is limited by the phenomenon of local shattering of the highly brittle crosslinked epoxy network. The addition of polyimide into epoxy systems has been attempted by many researchers to improve this situation by producing crack blunting via a fiber reinforcement mechanism but, as mentioned earlier, low percentages of polyimide due to solubility and various other processing difficulties [65, 74–81, 141] resulted in only limited testing for physical properties. Problems experienced in making bulk samples of epoxy-polyimide have resulted in only a few of the reports providing some limited physical property values [40, 74, 142].

6.1
Thermal Stability

The degradation of DGEBA-PMDA epoxy starts at about 310 °C and the breakdown of the material is reflected in a discontinuity in some of the DMA plots at around 315 °C. The thermal stability of the DGEBA-ODA/PMDA PI composite systems can be seen in Fig. 12. From this, it is seen that the 10 wt% loss temper-

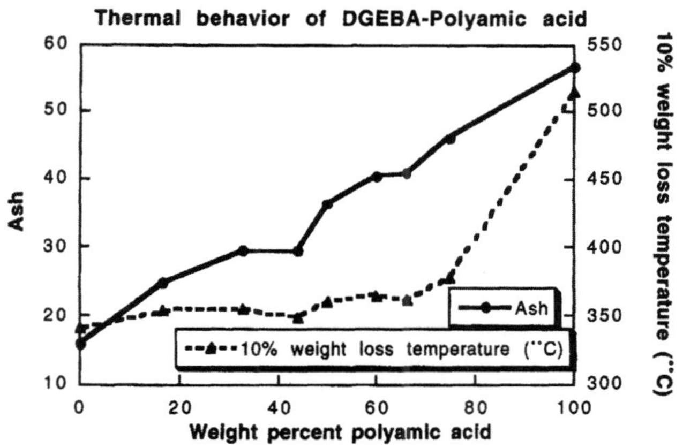

Fig. 12. The thermal behavior of DGEBA-PAA (ODA/PMDA) – *dashed line* 10% weight loss, *solid line* ash content at 600 °C

ature of the composites is not significantly increased with incorporation of the polyimide. This is due primarily to the low thermal stability of the epoxy component. The systems tested all had more than 10 wt% epoxy, so, it is this portion of the mixture that is decomposing at the low temperatures. The amount of ash in the systems has been noted as a indicator of thermal resistance [143] but from these results it is clear that this is not a good measure of thermal resistance of epoxy -polyimide composites.

6.2
Adhesion Properties of Epoxy Polyimide IPNs

This material could be used as an adhesive in microelectronics packaging applications, where the T_g of the material might be exceeded for a short period of time. This is typical in wire bonding applications, connecting the chip package to the circuit board, where the application of solder exposes the material to temperatures in excess of 260 °C for a few seconds or less. But the use of epoxies with T_gs near the soldering temperature, coupled with the high temperature stability brought about by the PI, make epoxy-polyimide systems particularly attractive.

Adhesion studies of epoxy resins modified with high modulus and high glass transition temperature thermoplastics have shown adhesion can reach or even exceed that of the unmodified resin. The use of flexible polyamides or flexible epoxides resulted in shear strength increases in epoxy systems employed by Cunliffe et al. [144], polyethersulfones [18, 145], polyetherimides [109, 146, 147], and polyetherketones [148–150].

Requirements of adhesives vary with the application but most epoxy adhesives have the following properties [151]:

1. ability to form self standing films (with or without pre-curing);
2. flow during lamination (fabrication);
3. high temperature resistance;
4. high peel strength;
5. low dielectric;
6. high T_g.

The generally accepted mechanisms of polymer-polymer adhesion include: (1) mechanical interlocking theory, (2) theories based on surface energies, wetting and adsorption, (3) diffusion theory, (4) electronic or electrostatic interaction, and (5) chemical bonding. The diffusion of polymer molecules across the interface allowing adhesive and adherend chains to entangle seems to be generally accepted as playing the major role in polymer-polymer adhesion [152, 153]. Epoxy adhesives, due to their inherent brittleness, have poor resistance to peel stress, the most common method of adhesion testing. Toughening of the epoxy by the introduction of a tough polyimide additive could and does lead to a material less prone to crack propagation and a better adhesive.

Crosslink density has been found to have a direct correlation to the T_g and the adhesive properties of the final material [154, 155]. In 180° peel tests conducted by the author, incorporation of epoxy into the polyimide has increased the ad-

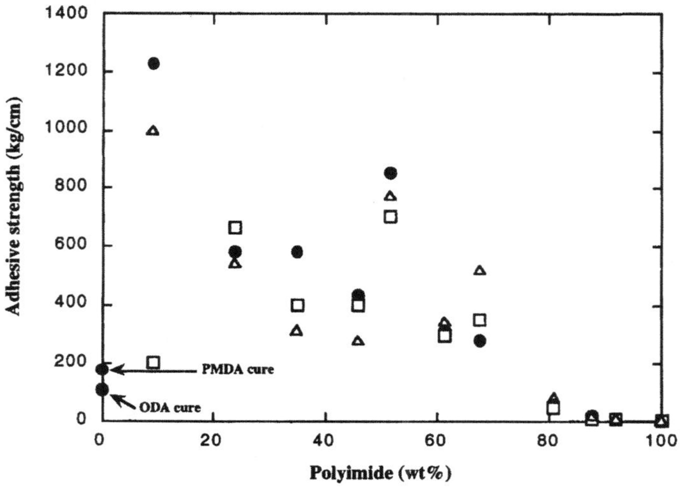

Fig. 13. Results of 180° peel adhesion tests for epoxy cured with polyamic acid systems (*circles*=cure for 1 h at 125 °C, 2 h at 250 °C, *squares*=cure for 1 h at 250 °C, *triangles*=low molecular weight PAA, cure for 1 h at 125 °C, 2 h at 250 °C)

hesion properties of the material. In the tests of adhesion strength of the DGEBA-PAA materials, typically the higher the level of epoxy content the more cohesive failure in the specimen was noted. In adhesion tests of DGEBA-PAA-PMDA systems, strengths of 1380 gm/cm were found for the 50/38/12 system. The results of the DGEBA-PAA adhesion experiments can be seen in Fig. 13.

Although the mechanisms of polyimide/metal adhesion remain to be fundamentally elucidated, it is generally accepted that the interfacial diffusion of metallic entities into the polyamic acid plays a key role at the interface [156–158]. Two main theories have been reported explaining the adhesion of the PI/metal bond: chemical and mechanical bonding [159]. Initial work emphasized mechanical bonding and most efforts were dedicated to the physical roughening of the substrate by different abrasive methods as well as chemical treatments in order to improve metal to polyimide adhesion by increasing the metal surface area [156, 160–164].

6.3 Mechanical and Dielectric Properties

In measurements of physical properties of epoxy resins – and the epoxy-polyimide system will be included in this general categorization – the method of failure is of considerable interest. Gross flaws in the epoxy structure act as initiation points for cracks which typically propagate by the breaking crosslinks. Therefore the level of crosslinking and cure will have a significant influence on the physical properties determined. The incorporation of additives that typically make the system flexible and affect the crosslinking density of the systems have a large influence on the mechanical strength.

Another important factor that would be important in many of the applications where epoxy-polyimide could be used is the dielectric constant. The high dielectric values of epoxy, in the range 3.5–10, result from the many hydroxyl groups in cured epoxy resin and preclude its use in electrically demanding applications. Polyimide with its excellent physical properties of high temperature stability and low dielectric constant make it an ideal candidate to replace the glass reinforcement in epoxy printed circuit boards.

Polyimides have low dielectric constants ranging from 2.7 to 3.5, have a good overall balance of thermal, mechanical, and electrical properties, and are frequently used as interlayer dielectrics. Polyimide interlayers allow for high chip packing densities and fast operation speeds in high performance computers and telecommunications. These applications place several demands on the materials such as high adhesion, low moisture absorption, strong adhesion to metal, chemical inertness, good mechanical properties, thermal stability reliability and easy film application [165, 166].

However, in unpublished works in the author's laboratory the use of polyamic acids to cure epoxy resulted in no decrease in dielectric constants of the epoxies examined. This lack of improvement was attributed to the presence of a large number of unreacted carboxyl groups and the presence of large quantities of hydroxyl groups from the incompletely crosslinked epoxy network.

7 Future Studies

The author notes many interesting avenues of future research that could be carried out to see if the potential for epoxy-polyimide composites can be realized and the physical phenomena that they exhibit could be explained.

First, the number of epoxies and the use of higher polyimide contents via the use of the THF-MeOH solvent system should be investigated further. This will open up the possibility of manufacturing bulk components of epoxy-polyimide. Elucidation of the mechanisms responsible for the solubility of the polyamic acid in non-traditional solvents such as tetrahydrofuran/methanol could open the door for the development of other solvent systems for seemingly intractable polymer systems.

The placement of reactive groups on the polyamic acid precursors that do not participate in the closure of the imide ring but could participate in the ring opening of the epoxy should be attempted. This would enable the polyimide to have chemically bonds with the epoxy network in a much higher density. The placement of large side groups such as cardo groups or flexible linkages in the polyimide backbone could be easily accomplished by the judicious use of the appropriate amines. Perhaps the ability of these large side groups to anchor physically the polyimide in the cured epoxy network would influence pullout of polyimide molecules upon fracture.

Analysis of the fracture surfaces of epoxy-polyimide via atomic force microscopic (AFM) analysis could reveal whether or not individual polyimide molecules are acting as fiber reinforcement.

The investigation of the gas permeation characteristics of epoxy-polyimide would also be interesting in noting the packing efficiency of the polyimide molecules and the effect of immobilization in an epoxy network would have on the permeation characteristics of various gases. The extent of epoxy crosslinking would also have an effect these properties of the composites.

Cure of epoxy with combinations of monomeric curing agents and the use of polymers containing reactive groups to cure epoxy systems is rarely touched upon in the literature and could be investigated much further. Amides, polyesters, polyethers etc. used to cure the epoxy network could lead to the creation of materials having interesting and heretofore unfathomed properties. The fundamental kinetics of the reactive polymer epoxy reaction is an area of interest as well.

The door into the fabrication of polyimide-epoxy compostites has been opened by the advent of the THF/MeOH solvent system – it is up to future reserachers to enter and explore this interesting corner of the polyimide labyrinth.

8
References

1. Seymour RB, Carraher CE (1984) Structure property relationships in polymers. Plenum Press, New York
2. Lee H, Neville K (eds) (1967) Handbook of epoxy resins. McGraw-Hill, New York
3. Patel RT, Patel VS (1994) Phosphorus, Sulfur and Silicon 89:113
4. Titier C, Rozenberg B, Taha M (1995) J Polym Sci Pt A, Poly Chem 33:175
5. Kim BS, Inoue T (1995) Polymer 36(10):1985
6. Yamanaka K, Takagi Y, Inoue T (1989) Polymer 30:1839
7. Yamanaka K, Inoue T (1985) J Mater Sci 25:241
8. Butta E, Marchetti A, Larreri A (1986) Poly Sci & Eng 26(1):63
9. McGarry FJ (1970) Proc R Soc London A319:59
10. Kunz SC, Sayre JA, Assink RA (1986) Polymer 23:1897
11. Takemura A, Tomita B, Mizumachi H (1985) J Appl Polym Sci 30:4031
12. Bucknall CB, Yoshii T (1978) Brit Poly J 10:53
13. Wang TT, Zupko HM (1981) J Appl Polym Sci 26:2391
14. Bartlett P, Pascault JP, Sautereau H (1985) J Appl Polym Sci 30:2955
15. Kinloch AJ, Gilbert DG, Shaw SJ (1986) J Mater Sci 21:1051
16. Yee AF, Pearson RA (1986) J Mater Sci 21:2462
17. Bussi P, Ishida H (1994) J Appl Polym Sci 53:441
18. Kim BS, Chiba T, Inoue T (1993) Polymer 34:2809
19. DeNograro FF, Mondragon I (1995) J Appl Poly Sci 56:177
20. Gupta VB (1984) J Macromol Sci Phys B23(4/6):435
21. Fizel MC, Hawley MC (1995) J Polym Sci Part A, Poly Chem 33:673
22. Bistrup SA, Simpson JO (1995) J Polym Sci Part B, Poly Phys 33:43
23. Palmese GR, McCullough RL (1992) J Appl Poly Sci 46:1863
24. Iannacchione G, VonMeerwal E (1991) J Polym Sci Part B, Poly Phys 29:659
25. Vallo CI, Williams RJJ (1991) J Polym Sci Part B, Poly Phys 29:1503
26. Meyer F, Sanz G, Mijovic (1995) Polymer 36(7):1407
27. Mangion MBM, Wang M, Johari JP (1995) J Polym Sci Poly Phys 30:445

28. Abadie MJM, Sillion B (eds.) (1991) Polyimides and other high temperature polymers. Elsevier, Amsterdam
29. Bessonov MI, Koton MM, Kuryavtsev VV, Laius LA (1987) Polyimides – thermally stable polymers. Plenum Press, New York
30. Feger C, Khojasteh MM, McGrath JE (eds) (1989) Polyimides: materials, chemistry and characterization. Elsevier, Amsterdam
31. Wever WD, Gupta MR (eds) (1987) Recent advances in polyimide science and technology. Mid-Hudson SPE, New York
32. Mitall, KL (ed) (1984) Polyimides: synthesis, characterization and applications, Vols 1 and 2. Plenum Press, New York
33. Bessonov MI, Zubkov VA (1993) Polyamic acids and polyimides. CRC Press, Boca Raton
34. Martinez PA, Cadiz V, Mantecon A, Serra A (1985) Die Ange Makro Chem 133:97
35. Martinez PA, Cadiz V, Mantecon A, Serra A (1987) Die Ange Makro Chem 148:149
36. Soler H, Cadiz V, Serra A (1987) Die Ange Makro Chem 148:152
37. Serra A, Cadiz V, Mantecon A (1987) Die Ange Makro Chem 148:155
38. Cadiz V, Mantecon A, Serra A, Thepaut C (1992) Die Ange Makro Chem 195:129
39. Roig A, Serra A, Cadiz V, Mantecon A (1992) Die Ange Makro Chem 199:75
40. Jang J, Lee W (1994) Polymer J 26:513
41. Seo Y, Hong SM, Hwang SS, Park TS, Kim KU, Lee S (1995) Polymer J 36:515
42. Eibl R, Hawthorne G, Hodgkin J, Jackson M, Loder J, Morton T (1993) Advanced Composites '93, International Conf on Adv Comp Materials. Minerals, Metals & Materials Soc
43. Chin WK, Shau MD, Tsai WC (1995) J Poly Sci Pt A 33:373
44. Yu HS, Yamashita T, Horie K (1996) Poly J 28:703
45. Yu HS, Yamashita T, Horie K (1996) Macromol 29:1144
46. Hay JN, Woodfine B, Davies M (1996) High Perform Poly 8(1):35
47. Sefton MS, McGrail PT, Peacock JA, Crick RA, Davies M, Almen G (1987) Polymers for composite. PRI, London
48. Takayanagi M, Ogata T, Morikawa M, Kai T (1980) J Macromol Sci Phys B17:591
49. Hwang WF, Wiff DR, Verschoore C, Price GE, Helminiak TE, Adams WW (1983) Polym Eng Sci 23:784
50. Kumar S, Wang CS (1989) Polym Commun 29:355
51. Lee CYC, Swaitkeiwicz J, Prassad PN, Mehta R, Bai SJ (1991) Polymer 32:1195
52. Roberts MF, Jenekhe SA (1994) Chem Mater 6:135
53. Flory PJ (1978) Macromolecules 11:1138
54. Flory PJ (1941) J Chem Phys 9:660
55. Paul DR, Sperling LH (1986) Multicomponent polymer materials, advances in chemistry series, 211, chap 5. ACS, Washington D.C.
56. Chang K-Y, Chang H-M, Lee Y-D (1994) J Poly Sci Part A Poly Chem 32:2629
57. Lipatov YS (1990) J Macromol Sci Rev C30:209
58. Akay M, Rollins SN (1993) Polymer 34(9):1865
59. Zhou P, Frisch HL, Rogovina L, Mararova L, Zhdanov A, Serfeienko N (1993) J Polym Sci Pt A: Polym Chem 31:2481
60. Hourston DJ, Huson MG (1992) J Appl Polym Sci 45:1753
61. Lipatov YS, Rosovitsky VF, Babkina NV (1993) Polymer 34(22):4697
62. Lin M-S, Chang R-J, Yang T, Shih Y-F (1995) J Appl Polym Sci 55:1607
63. Allen F, Bowden MJ, Blundell DJ, Hutchinson FG, Jeffs GM, Vyvoda J (1973) Polymer 14:597
64. Jia DM, You BW, Wang MZ (1988) Int Polym Process 3(4):205
65. Morin A, Djomo H, Meyer GC (1983) Polym Eng Sci 24(11):1415
66. Kim BS, Lee DS, Kim SC (1986) Macromolecules 19:2589
67. Kim SK, Kim SC (1990) Polym Bull 23:141
68. Chang R-J, Yang T, Shih Y-F (1995) J Appl Polym Sci 55:1607
69. Musto P, Ragosta G, Scarinzi G (1993) Ange Makro Chem 204:153
70. Martuscelli E, Musto P, Ragosta G, Scarinzi G (1993) Ange Makro Chem 213:93

71. Di Liello V, Martuscelli E, Musto P, Ragosta G, Scarinzi G (1994) J Polym Sci Pt. B: Polym Phys 32:409
72. Iijima T, Tochimoto T, Tomoi M (1991) J Appl Polym Sci. 43:1685
73. Gaw K, Jikei M, Kakimoto M, Imai Y (1996) Reactive & Func Poly 30:85
74. Sillion B, Boilley N, Pascal T (1994) Polymer 35(3):558
75. Yu JW, Sung CS (1995) Macromolecules 28:2506
76. Chin WK, Shau MD, Tsai WC (1995) J Polym Sci Part A, Poly Chem 33:373
77. Patel HS, Shah VJ (1993) High Perf Poly 5:145
78. Patel HS, Shah VJ (1994) Macromolecular Reports, A31 (Suppl 5):545
79. Patel HS, Shah VJ (1995) J Mat Sci-Pure and Appl Chem A32(3):405
80. Patel HS, Shah VJ (1994) Interntl J Polym Mat 26:79
81. Patel HS, Shah VJ (1994) Bull Mater Sci 17(4):361
82. EchigoY, Iwaya Y, Tomioka I, Furukawa M, Okamoto S (1995) Macromolecules 28:3000
83. EchigoY, Iwaya Y, Tomioka I, Yamada H (1995) Macrolomecules 28:4861
84. EchigoY, Iwaya Y, Saito M, Tomioka I (1995) Macrolomecules 28:6684
85. Gaw K, Suzuki H, Jikei M, Kakimoto M, Imai Y (1997) Poly J 29:290
86. Kryauf D, Strehmel V, Fedke M (1993) Polymer 34(2):323
87. Fu JH, Schlup JR (1993) J Appl Poly Sci 49:219
88. Abbate M, Martuscelli E, Musto P, Ragosta G, Scarinzi G (1994) J Poly Sci Pt B. 32:395
89. Varley RJ, Heath GR, Hawthorne DG, Hodgkin JH (1995) Polymer 36:1347
90. Maruscelli E, Musto P, Ragosta G, Scarinzi G, Bertotti E (1994) J Poly Sci Pt. B 31:619
91. Scherzer T (1994) J Appl Poly Sci 51:491
92. Strehmel V, Scherzer T (1994) Eur Poly J 30:361
93. Smith RE, Larsen FN, Long CL (1984) J Appl Poly Sci 29:3713
94. Gallouedec F, Costa-Torro F, Laupretre F, Jasse B. (1993) J Appl Polym Sci 47:823
95. Kozielski KA, George GA, St. John NA, Billingham NC (1994) High Perform Polym 6:263
96. Mijovic J, Andjelix S, Winnie Yee CF, Bellucci F, Nicolais L (1995) Macromolecules 28:2797
97. Gupta A, Cizmecioglu M, Coulter D, Liang RH, Yavrouian A, Tsay FD, Moacanin J (1983) J Appl Poly Sci 28:1011
98. Shechter L, Wynstra J, Kukjy RP (1956) Ind Eng Chem 48:94
99. Dusek K, Bleha M (1977) J Polym Sci Polym Chem Ed 15:2393
100. Prime RB, Sacher E (1972) Polymer 13:455
101. Gaw KO (1996) PhD Thesis. Tokyo Institute of Tech, Tokyo
102. Coleman MM, Serman CJ, Bagwagar DE, Painer PC (1990) Polymer 31:1187
103. Flory, PJ (1953) Principles of polymer chemistry. Cornell University Press, Ithaca, New York
104. Krause S, Roman N (1965) J Poly Sci 3:1631
105. Makhija S, Pearce EM, Kwei TK, Liu F (1990) Polymer Eng & Sci 30:798
106. Ratzsch MJ (1994) Material Science-Pure and Appl Chem A31(10):1399
107. Bucknall CB, Partridge IK (1983) Polymer 24:639
108. Bucknall CB, Gilbert AH (1989) Polymer 30:213
109. Bucknall CB, Partridge IK (1986) Poly Eng & Sci 26:54
110. Gibbs JW (1928) Collected works, vol 1. Longmans, Green, NYC
111. Min B-G, Hodgkin JH, Stachurski ZH (1993) J Appl Polym Sci 48:1303
112. Kimura M, Porter RA (1984) J Polym Sci Polym Phys 22:1697
113. Siegmann A, Dagan A, Kenig S (1985) Polymer 26:1325
114. Paci M, Barone C, Magagnini PL (1987) J Polym Sci Polym Phys 25:1595
115. Weiss RA, Huh W, Nicolais L (1987) Polym Eng Sci 27:684
116. Levita G, DePetris S, Marchetti A, Lazzeri A (1991) J Mater Sci 26:2348
117. Pearson RA (1991) J Mater Sci 26:3828
118. Whang WF, Wiff DR, Benner CL, Helminiak TE (1983) Macromol. Sci Phys B22(2):231
119. Olabisi O, Robeson LM, Shaw MT (1979) Polymer-polymer miscibility. Academic Press, London

120. Gillham JK (1986) Polymer Eng & Sci 26:142
121. Couchman PR, Karasz FE (1978) Macromolecules 11:117
122. Couchman PR (1978) Macromolecules 11:1156
123. Nandi AK, Mandal BM, Bhattacharyya SN, Roy SK (1986) Poly Comm 27:151
124. Small PA (1953) J Appl Chem 3:71
125. Fedors RF (1974) Polym Eng Sci 14:147
126. Kelley FN, Bueche F (1961) J Poly Sci 50:549
127. Gordon M, Taylor JS (1952) J Appl Chem 2:493
128. Gibbs J, Dimarzio E (1952) J Poly Sci 40:121
129. Zukas WX (1994) J Appl Poly Sci 53:429
130. Horie K, Hiura H, Sawada M, Mita I, Kambe H (1970) J Poly Sci Pt A-1 8:1357
131. Wu C-S (1992) J Mat Sci 27:2952
132. Gan SN, Burfeild DR (1989) Polymer 60:1905
133. Fry CG, Lind AC (1990) New Polym Mater 2:235
134. Wedgewood AR, Seferis JC (1981) Polymer 22:966
135. Sanz G, Garmendia J, Andres MA, Mondragon I (1995) J Appl Poly Sci 55:75
136. Latha PB, Adhinarayanan K, Ramaswamy R (1994) Int J Adhesion Adhesives 14:57
137. Tanaka Y, Kaiuchi H (1963) J Appl Polymer Sci 7:1063
138. Kohli A, Chung N, Weiss RA (1989) Polym Eng Sci 29:573
139. Skovby M, Kops J, Weiss RA (1991) Polym Eng Sci 31:954
140. Beery D, Kenig S, Siegmann A (1991) Polym Eng Sci 31:451
141. Kimoto M, Yoshioka Y, Mizutani K, Mitoh M (1995) Proc Jpn Adhs Conf 33:9
142. Boilley N, Pascal T, Sillion B (1991) In: Abadie MJM, Sillion B (eds.) Polyimides and other high temperature polymers. Elsevier, Amsterdam, p 319
143. Van Krevelen DW (1975) Polymer 16:615,
144. Cunliffe AV, Huglin MB, Pearce PJ, Richards DH (1975) Polymer 16:659
145. Yamanaka K, Inoue T (1989) Polymer 30:662
146. Kinloch AJ, Yuen ML, Jenkins SD (1994) J Mater Sci 29:3781
147. Raghava RS (1987) J Polym Sci, Polym Phys 25:1017
148. Hourston DJ, Lane JM (1992) Polymer 33:1379
149. Iijima T, Tochimoto T, Tomoi M (1991) J Appl Polym Sci 43:463
150. Cerere JA, McGrath JE (1986) Polym Prepre 27:299
151. Miura O (1995) Adhesives in Multiple Chip Packages. ACS Symposium, Pacifichem'95, Honolulu
152. Brown HR, Yang ACM, Tussel TP, Volksen W (1988) Polymer 29:1807
153. Ellul DM, Gent AN (1984) J Polym Sci Polym Phys Ed 22:1953
154. Levita G, DePetris S, Marchetti A, Lazzeri A (1991) J Mater Sci 26:2348
155. Pearson RA (1991) J Mater Sci 26:3828
156. Pappas DL, Cuomo JJ (1991) J Vac Sci Technol A9(5):2704
157. Buchwalter LP (1990) J Adhes Sci Technol 4:697
158. Lindsey WB (1986) U.S. Pat 3,361,589
159. Chen K-M, Ho S-M, Wang T-H, King J-S, Chang W-C, Cheng R-P, Hung AJ (1992) Appl Polym Sci 45:947
160. Knorre H, Meyer-Simon E (1972) U.S. Pat 3,702,285
161. Hermer J (1973) U.S. Pat 3,770,528
162. DeAngelo MA (1974) U.S. Pat 3,791,848
163. Yates CB Wolski AM (1974) U.S. Pat 3,857,681
164. Tromp RM, Legoues F, Ho PS (1985) J Vac Sci Technol A3:782
165. Chen K-M, Ho S-M, Wang T-H, King J-S, Chang W-C, Cheng R-P, Hung A (1992) J Appl Polym Sci 45:947
166. Hu D-C, Chen H-C (1992) J Mater Sci 27:5262

Received: March 1998

Thermosetting Oligomers Containing Maleimides and Nadimides End-Groups

Pierre Mison, Bernard Sillion

LMOPS UPR 9031 CNRS, BP 24, F-69390 Vernaison, France
E-mail: lmops69@imaginet.fr

In the field of high thermomechanical performance polymers, linear and thermosetting systems offer complementary properties. Among the thermosetting materials, BMIs and BNIs have been extensively studied and are now commercially available. In this chapter, firstly the main preparation and characterization methods are reviewed, and then the chemistry of the polymerization processes is discussed for both families. For the BMIs, due to the electrophilic character of their double bond, different polymerization pathways have been published, which is not the case for BNIs. Special attention has been paid to thermal polymerization which has already been used in industrial achievements; however, on the other hand, the structure of these materials has been considered for the purpose of establishing relationships between processability, stability and thermomechanical properties.

Advances in Polymer Science, Vol.140

List of Abbreviations

BBN	Benzhydrol bisnadimide
BCI(1)	Biscitraconimide(s)
BHTDA	3,3',4,4'-Benzhydroltetracarboxylic dianhydride
BMI(s)	Bismaleimide(s)
BNI(s)	Bisnadimide(s)
BP	British Petroleum
BTDA	3,3',4,4'-Benzophenonetetracarboxylic dianhydride
CA	Cycloadduct
CHP	*N*-Cyclohexyl-2-pyrrolidone
CPD	Cyclopentadiene
CPMAS	Cross Polarization Magic Angle Spinning
DA	Diels–Alder
DABCO	1,4-Diazabicyclo[2.2.2]octane
DMAC	*N,N*-Dimethylacetamide
DMF	*N,N*-Dimethylformamide
DSC	Differential scanning calorimetry
E_a(s)	Activation energy(ies)
ESR	Electron spin resonance
FTIR	Fourier transform infrared
HFDA	3,3'-[1,1,1,3,3,3-Hexafluoro-2,2-propylidene]diphthalic anhydride
HFDE	Dimethyl ester of HFDA

HPLC	High performance liquid chromatography
IR	Infrared
LSIMS	Liquid secondary ion mass spectrometry
LARC TPI	Langley Research Center Thermoplastic Polyimide
MA	Maleic anhydride
MAL	Maleimide
MDA	4,4'-Methylene-1,1'-dianiline
NA	Nadic anhydride
NAD	Nadimide
NASA	National Aeronautics and Space Administration (USA)
NMP	Nuclear magnetic resonance
PDA	Phenylenediamine, m: *meta*, p: *para*
PES	Polyethersulfone
PMR	Polymerization of monomeric reactants
rDA	Reverse Diels–Alder
Semi-IPN	Semi-interpenetrated network
THPI	1,2,3,6-Tetrahydrophthalimide
Da	Daltons
G_{1c}	Strain energy release rate
K_{1c}	Stress intensity factor
$\bar{M}_n$	Number-average molecular mass
Mp	Melting point
ppm	Parts per million
T_g	Glass transition temperature
Tm	Maximum of the temperature of the melting endotherm curve (DSC)
Tp	Onset temperature of the polymerization of the exotherm curve (DSC)
ΔH	Reaction enthalpy

1 Introduction

The nadimide and maleimide end-capped oligomers belong to the family of thermosetting telechelic oligomer precursors of thermostable networks shown in Fig. 1. These products were developed in order to fulfill the requirements of the aerospace industry in the domain of high performance adhesives and matrices for laminates.

The thermomechanical properties of an organic material mainly depend on two factors. Firstly the molecular relaxations (crystalline melting point and glass transitions) which determine the temperature upper limit for applications, and secondly the chemical nature of the backbone which is responsible for the stability in a harsh environment.

Aromatic and heterocyclic polymers exhibit high glass transitions and excellent chemical stabilities. They were developed for high-temperature applica-

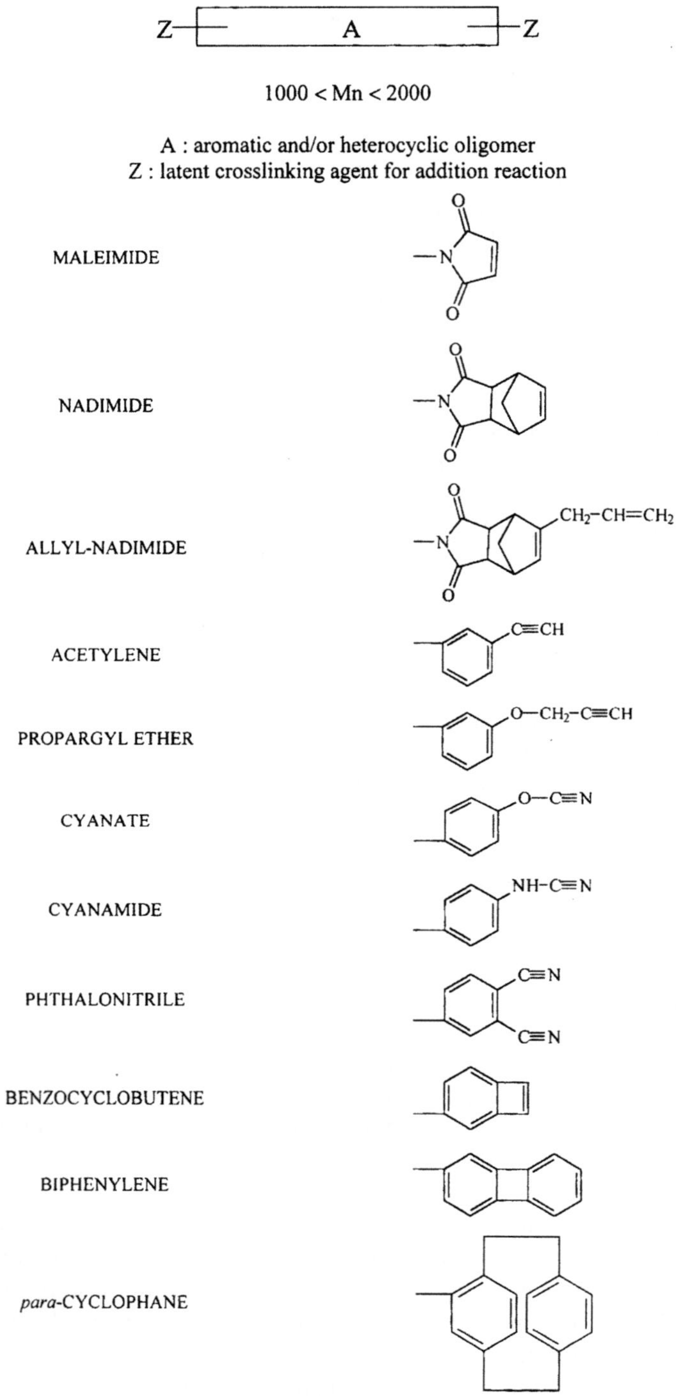

Fig. 1. Main types of end-groups for thermosetting oligomers

tions, but such high molecular weight polymers are very difficult to process due to the lack of solubility and the very high viscosity above their T_g.

It is well known that by reducing the molecular weight of a polymer its T_g decreases and its solubility increases. As a consequence the processability of an aromatic or heterocyclic oligomer is better than that of the corresponding high molecular weight polymer. The mechanical properties are obtained after crosslinking of the end-capping group.

This concept was introduced during the 1960s and 1970s and the first studied families were the maleimide and nadimide terminated oligomers due to an interesting compromise between processability, thermomechanical properties and cost. Right now both of these materials are used in industry for aerospace and electronic applications. Several reviews or book chapters have been devoted to these types of oligomers [1–5].

In this chapter a comparative study of the bismaleimide and bisnadimide oligomers will be presented with the following items:

- oligomer syntheses,
- characterization methods for reactants and polymerization processes,
- chemistry of bismaleimides and bisnadimides
- structures and properties of the networks obtained by thermal crosslinking.

2 Oligomer Syntheses

2.1 Bismaleimides (BMIs)

The BMIs are prepared according to a two-step process based on the reaction of a diamine with maleic anhydride. The first reaction is usually performed at room temperature in an aromatic, or chlorinated, or aprotic solvent. This fast and exothermic reaction leads to the monoamide (so called maleamic acid) of maleic acid. However formation of the monoamide of fumaric acid (*trans*-isomer) was observed [6] and the formation of the *trans*-isomer which does not cyclize, has to be minimized by working at low temperature.

Figure 2 shows the chemistry of the maleic anhydride condensation with an aromatic amine followed by the cyclization to the maleimide. The cyclization can be performed by heating without isolation of the amic acid intermediate and it has been suggested [7] that the thermal cyclization is improved using a solvent-water azeotropic distillation system based on DMAC and CHP at 130 °C. This process seems interesting for the end-capping of oligomers with a number-average molecular weight ranging between 2,000 and 10,000 g mol^{-1}. For the preparation of low molecular weight aromatic BMIs the chemical dehydration with acetic anhydride and a basic catalyst like sodium acetate is generally used. The formation of isoimides as intermediates, and of acetamide of the aromatic diamine as side product were evidenced [8].

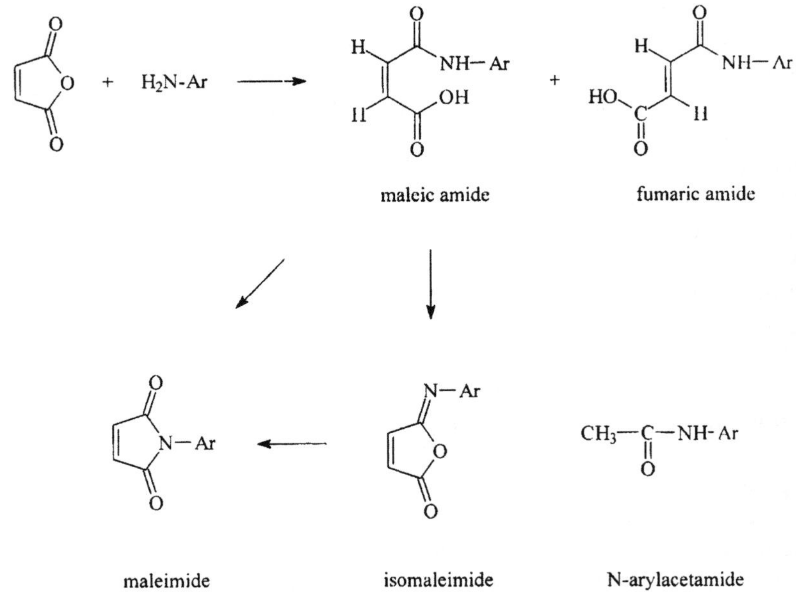

Fig. 2. Reactions which take place during the maleimide synthesis

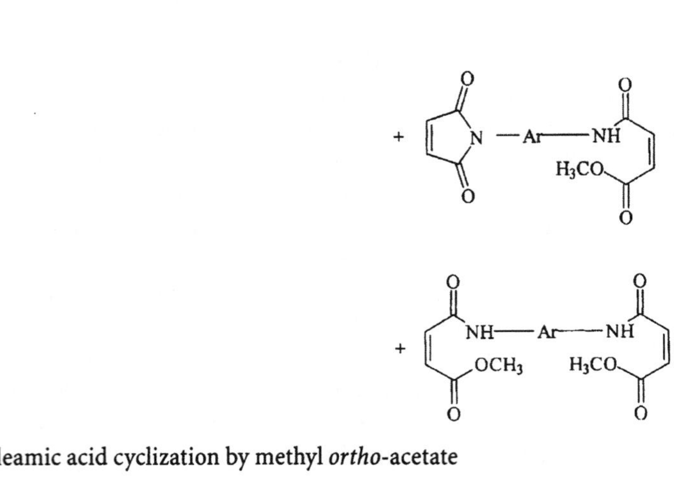

Fig. 3. Maleamic acid cyclization by methyl *ortho*-acetate

Some organic bases like tertiary amines can also catalyze the amide acid cyclization [6,9], and the nature of the tertiary amine allows the control of the proportions of imide-isoimide-acetamide in the reaction product. For example the reaction of MA with MDA using sodium acetate as catalyst give a BMI with a crystalline melting point at 160 °C. If the condensation is carried out with DABCO, an amorphous mixture containing BMI, monomaleimide-mono isoimide, and monomaleimide-monoacetamide is obtained. This amorphous material called Desbimid is more processable than the pure BMI but exhibits similar thermomechanical properties [10].

Another type of dehydrating agent, the ortho esters, has also been used for the preparation of processable BMIs. A DMF solution of MDA bismaleamic acid was treated with methyl orthoacetate and triethylamine. A mixture of three products shown in Fig. 3 was obtained with a softening point ranging from 98 to 106 °C [11].

2.2 Bisnadimides (BNIs)

Mononadimides, bisnadimides and nadimides end-capped oligomers are generally prepared in a one-stage process. The starting materials are dissolved in a polar solvent (NMP, DMF, diglyme,...). Intermediate amic acids are formed by heat-

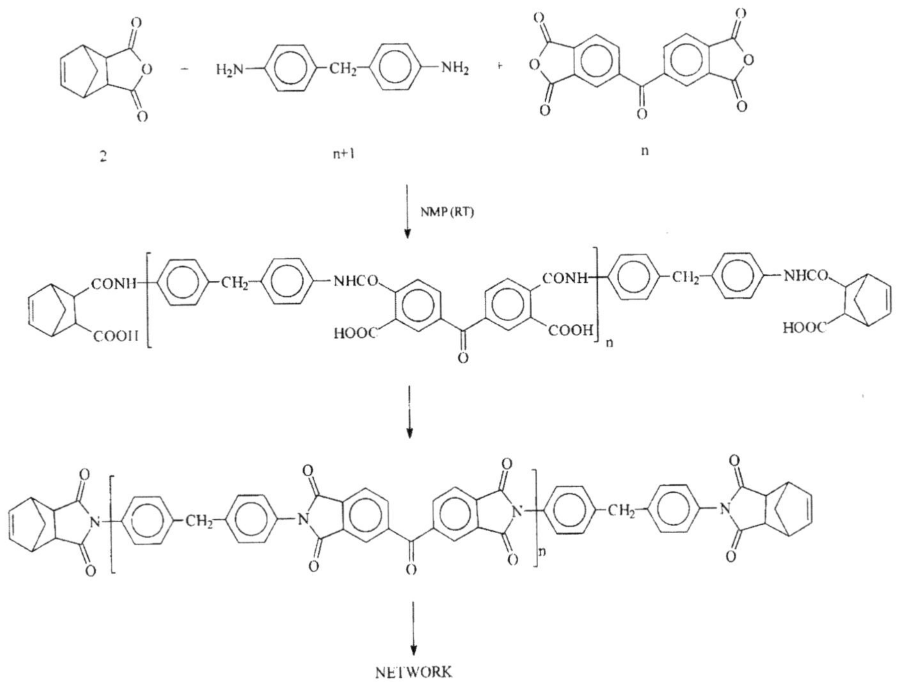

Fig. 4. Chemistry of the first bisnadimide end-capped oligomer (theoretical n=1.67; $\bar{M}_n$= 1300 Da)

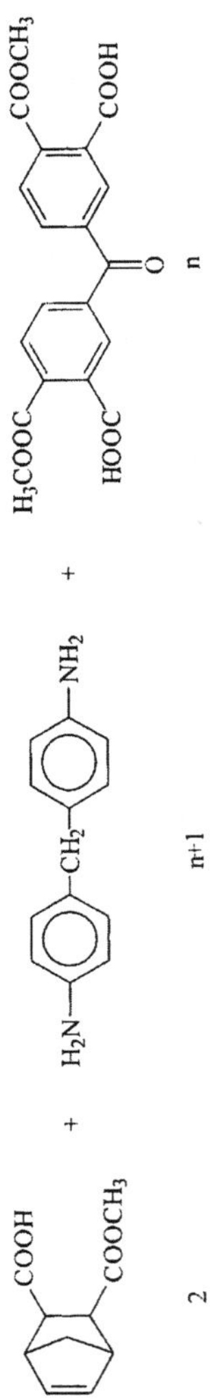

- Alcoholic solution
- Prepeg preparation with carbon clothe

Fig. 5. PMR-15 chemistry (theoretical n=2.087)

ing the solution at 80 °C. The imidization is then accomplished by raising the temperature to 160 °C and maintaining it for some hours (≈5). Water formed can be eliminated by various techniques (use of a Dean-Stark apparatus, azeotropic distillation with toluene...).

The first nadimide end-capped oligomers were prepared using the unbalanced classical reaction of a dianhydride (BTDA) with an excess of diamine (MDA) and a quantity of nadic anhydride (NA) corresponding to the diamine excess (Fig. 4). The reaction was performed in NMP with a molecular ratio calculated in order to get a M_n of 1300 g mol^{-1} [12]. However the amic acid intermediate is not stable enough to allow a long term storage of the NMP solution

[13] and on the other hand the fully cyclized BNI shows a very high softening point leading to a narrow processability window.

These drawbacks explain the development of a new concept introduced by a NASA team. According to this concept "polymerization of monomeric reactants" (PMR), a bisortho ester acid replaces the dianhydride and the monomethyl ester of the nadic acid is used instead of the nadic anhydride as shown in Fig. 5. These esters and the diamine are dissolved in methanol (or in a methanol/isopropanol mixture) in a molar ratio calculated to get the desired theoretical molecular weight [14]. These mixtures of reactants are currently used for the preparation of composites and adhesives [15a,b].

3 Characterization Methods for Reactants and Polymerization Processes

3.1 Introduction

The BMI and BNI formulations are in fact mixtures of products obtained either by synthesis or by blending. In the majority of cases, they are prepared by a method including unbalanced polycondensation followed by end-capping with maleic or nadic anhydride. So, analytical characterizations must take into account determination of molecular weight, polydispersity, functionality and characterization of side reactions which appear during the preparation of the starting materials. On the other hand, attention has been paid to the polymerization in the solid state in order to determine the extent of polymerization after gelation, the structure of the network and the degradation mechanisms.

3.2 BMIs

When the end-capping is carried out in two separate steps the titration of the acidic groups give an accurate view of the oligomer molecular weights. This method also allows control of the formation of the imide by following the disappearance of the amic acid. Various titration techniques are available but the most common is to use a tetralkylammonium hydroxide in methanol, the equivalent point being determined by potentiometry [6,16].

FTIR spectroscopy is widely used to monitor the dehydration of amic acids and the maleimide polymerization. The characteristic wave numbers which can be used are given in Table 1 [17].

The disappearance of the NH absorption, the appearance of in-plane and out-of-plane carbonyl vibrations, and the appearance of the C-N-C band at 1390 cm^{1} are the most accurate signals which have to be considered. During the maleic end-capping of amines terminated at polyether sulfone, the maleimide formation was characterized by the variation of the ratio of the absorbance at

Table 1. FTIR wavenumbers characteristic of amic acid and imide groups [17]

Wavenumber (cm^{-1})		Assignment
Amic acid	Imide	
3300/3200	–	ν (N-H)
–	1770	ν (C=O) in-plane
–	1730/1710	ν (C=O) out-of-plane
1560/1530	–	δ (N-H)
–	1390	ν (C-N-C)
1270/1250	–	δ (N-H)
–	740/700	δ (O=C-N)
650/630	–	δ (N-H)
Succinimide	Maleimide	
1180	1150	ν (C-N-C)

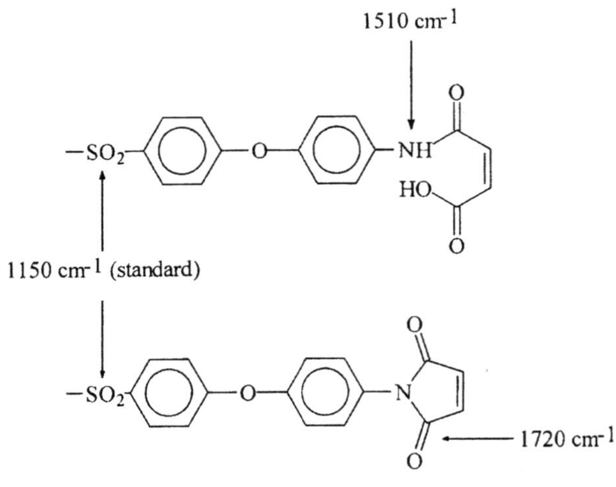

Fig. 6. Example of IR monitoring for maleimide end-capped polysulfone [6]

1720 cm^{-1} (maleimide) over the one at 1510 cm^{-1} (maleamic acid). The base line was the band centered at 1850 cm^{-1} and the absorption at 1150 cm^{-1} ($-SO_2$) was used as internal standard [6] (Fig. 6).

During the solid-state polymerization the maleimide disappears as the substituted succinimide forms as shown in Fig. 7. An aromatic absorption at 1515 cm^{-1} similar for both the maleimide and crosslinked succinimide network was used as an internal standard. The disappearance of the absorption band located at 1150 cm^{-1} (ν C-N-C) plotted as a function of the increase of the corre-

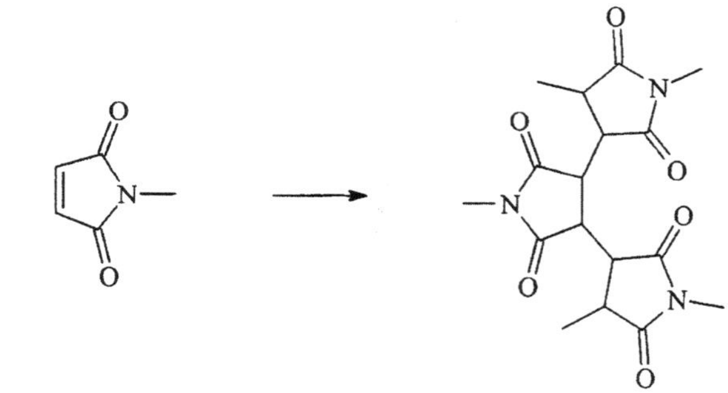

Fig. 7. Solid-state FTIR monitoring of the maleimide polymerization

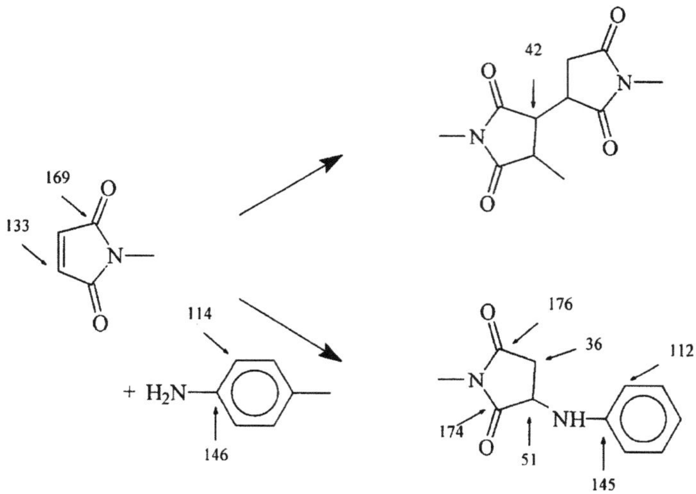

Fig. 8. Maleimide polymerization: nucleophilic amine addition vs. radical polymerization, ^{13}C NMR characterization (chemical shifts in ppm)

sponding succinimide absorption (ν C-N-C) at $1180\ cm^{-1}$ allows the kinetic determinations [17].

However, FTIR spectroscopy is not accurate enough to differentiate between the succinimide formation either by a nucleophilic addition of an aromatic diamine on the maleimide or by radical crosslinking. NMR spectroscopy offers a possibility to analyze the two different mechanisms and Fig. 8 shows the ^{13}C chemical shifts of different carbon atoms as a function of the product structures [17].

3.3
BNIs

Model compounds (mono- and bisnadimides) as well as nadimide end-capped oligomers have been extensively studied. In this section we will examine first the characterization of BNIs, then the monitoring of their thermal polymerization and finally the characterization of polymers.

3.3.1
Characterization of BNIs

The *endo*-nadimide isomer is obtained if the preparation temperature does not exceed 160 °C. The use of higher temperatures generally induces the partial isomerization of the *endo*-isomer to the *exo* one ; the product obtained in this case is a mixture of both isomers.

The imide formation can be either controlled by IR or NMR spectroscopy [18–23]. The formation of nadimide moieties is easily detected through specific absorption (Table 2), depending on the isomer isolated. IR and ^{1}H NMR analyses also allow detection of whether some uncyclized amic acids remain in the reaction medium by the presence of:

- strong absorptions at 3400 cm^{-1} (OH) and 1730 cm^{-1} (CO) in the IR spectra, and
- resonance signals at around 10–12 ppm (COOH), 9 ppm (NH) and 1.3 ppm (H-7,7') in ^{1}H NMR spectra.

Due to the wide use of PMR-15, a lot of work has been devoted to the characterization of nadimide end-capped oligomers. Two aspects have been considered. First the characterization of the imide moieties, which is conducted by spectroscopic measurements [24–28]. The second aspect is the determination of the oligoimide composition and of the average molecular weight ; this point has been largely investigated by chromatographic techniques (GPC) and has already been reviewed [15c]. More accurate results (not depending on standardization)

Table 2. Selected IR wavenumber absorptions (cm^{-1}) and NMR resonances (ppm) of *endo*-*exo* nadimide isomers

Spectroscopy	endo	exo	Assignment	Ref
IR	840	780	out-of-plane=CH conjugated with CO	18
^{1}H NMR	3.5	2.8	H-2,3*	[19–23]
	6.2	6.4	H-5,6*	[19–23]
	1.6	1.4	H-7,7'*	[19–23]
^{13}C NMR	52	43	C-7*	19,20
	135	138	C-5,6*	23

*Numbering of the norbornene moieties.

Table 3. ^{1}H NMR chemical shifts (ppm from TMS) of methylene in PMR-15 heated at 220 °C [26]

δ CH_2	Assignment of methylene signals
4.10	CH_2 located between 2 BTDE entities for n≥2
4.06	CH_2 located between 1 BTDE and 1 NAD (*exo*) entity (n≥1)
4.03	CH_2 located between 1 BTDE and 1 NAD (*endo*) entity (n≥1)
4.05	*exo-exo* configuration of BNI of MDA (n=0)
4.00	*exo-endo* configuration of BNI of MDA (n=0)
3.97	*endo-endo* configuration of BNI of MDA (n=0)
≈3.9–3.8	constituents containing amic acid or amic ester units
≈3.8–3.7	constituents containing terminal amine unit(s)
3.7–3.6	constituents containing terminal amine units and amic acid or ester unit(s)
3.55	free MDA

were obtained through ^{1}H NMR determinations [26,29]; however, such a technique generally requires high performance NMR devices and a separated resonance for the end-groups and for the oligomers. Qualitative IR data are also available [30].

The ^{1}H and ^{13}C NMR resonances of the methylene group of MDA are good markers to follow this moiety during the oligomerization step and the beginning of the crosslinking one of nadic systems (Table 3).

Thus, the oligomerization can be followed accurately to obtain an entirely imidized material. Moreover, at this last stage, it is possible to determine the relative amounts of lowest polycondensation degrees (n=0, n=1 and n≥2). In this way, it was determined that n=0 (bisnadimide of MDA, 35–50%) and n=1 oligomers (15–25%) are the largest constituents of the oligoimide [26]. The same kind of observations were done with oligobenzhydrolimides [31,32]. ^{13}C NMR spectroscopy also allows qualitative determinations [24,25,27,28].

3.3.2
Monitoring of the BNI Polymerization (Crosslinking) Process

Polymer properties are very often dependent on the polymer preparation. So, a good monitoring of the polymerization process is the key step to obtaining good and reproducible materials. The extent of the polymerization can be controlled in different ways. IR is the most usual [27,30] but is not very accurate and requires the extraction of samples to analyze. Recently, an in situ monitoring of PMR-15 processing has been provided by means of frequency-dependent dielectric measurements [33,34]. This non-destructive technique allows the characterization of all the steps of the curing process and thus they can be optimized.

A lot of solid properties are easily determined so large amounts of data are available about mechanical, rheological, thermal and ageing behaviors of polynadimides [35–37]. These will be presented in Sects 5 and 6. It is worth mentioning that dynamic mechanical thermal analyses have also been used to follow the oligomerization of some PMRs [37].

Rheological analyses give information about the viscosity and gel point. Such analyses have been coupled with thermogravimetry and have been used to determine a good way to prepare thermostable foams [32].

The beginning of the polymerization reaction can be followed by NMR spectroscopy. Thus the nadimide polymerization advancement is easily monitored by the disappearance of the ethylenic proton and carbon resonances. However these criteria should be handled with care. For instance, the ethylene proton disappearance was particularly misleading in the case of the 2,3,4,6-tetrahydrophthalimide curing study [38]. An isomerization stage took place with the formation of a tetrasubstituted double bond. Moreover, an oxidation reduction pathway was also evidenced, which gave saturated and aromatic derivatives.

The cure of PMR-15 and its model compound 4,4'-methylene dianiline bisnadimide (MDA, BNI) has been studied by simultaneous reaction monitoring and evolved gas analysis (SIRMEGA) using a FTIR with a mercury-cadmium telluride detector. The system allows the observation of the variation in IR spectra correlated to the gas evolution during the curing. The data show that the cyclopentadiene evolution involves only minor modifications in the spectrum [39].

3.3.3
Characterization of BNI Polymers: Structure Analysis Techniques

The usual difficulties in structurally characterizing crosslinked materials are mainly related to their insolubility in the usual organic solvents. Solid-state FTIR spectroscopy is the most convenient analysis to perform; several acute techniques allow accurate measurements to be obtained for several kinds of polymers including nadimide end-capped oligomers [30]. Nevertheless, only fragmentary structural information is obtained; the attendance in the network of some functional groups can be evidenced. However, it is not possible to determine exactly the polymer microstructure by FTIR.

Solid-state ^{13}C NMR is becoming increasingly used. The technique using samples in natural ^{13}C abundance leads to results difficult to analyze. Although providing precious information, they are not conclusive towards the structure determinations. Techniques using ^{13}C selectively labelled samples (obtained from starting materials labelled at sites that are thought to be involved in the polymerization reaction) are more efficient. By using the possibility of obtaining subtraction spectra of labelled and unlabelled materials that have been polymerized identically, only the labelled sites are present in the spectra [40]. Applied to the nadimide end-capped polyimide [41], this kind of analysis gave good support for the determination of polymer structure. ESR allows only surface analyses. Some qualitative information can be obtained from DSC measurements about the crosslinking density.

Modern techniques of mass spectrometry allow the determination of some polymer characteristics. The main studies are devoted to mass determinations (no standardization necessary) of polyethyleneglycols, polystyrenes and polymethylmethacrylates. However, a liquid secondary ion mass spectrometry (LSIMS) study [42] on a thermal polymer of a mononadimide model compound permits an approach to the determination of the polymer microstructure and goes some way to understanding the polymerization mechanism.

4 Chemistry of BMIs and BNIs

The BMI double bond exhibits a high and versatile reactivity due to conjugation with carbonyl groups. For example, the literature is well documented about step-growth polymerization including nucleophilic addition, Diels–Alder polymerization (maleimide is a good dienophile) and radical or anionic double-bond polymerization.

The non-conjugated BNI double bond is less reactive. However, the polymerization mechanism has not up to now been definitely established although a radical process is postulated for the propagation. Recent approaches towards the structure determinations of crosslinked networks and, by the way, towards the best knowledge of polymerization mechanisms, were undertaken in order to obtain BMI and BNI polymers presenting improved properties. In this section particular attention will be paid to this point.

4.1 BMI step-growth Polymerization by Nucleophilic Addition

Step-growth polymerization can be performed by reaction of diamine or dithiol on bismaleimide (Fig. 9). The reaction is a nucleophilic attack on the electron-deficient double bond activated by the two electron-withdrawing adjacent carbonyl groups.

The reaction with thiols takes place in *meta*-cresol or in acetic acid containing DMF. The slightly acidic medium is needed to favor the protonation of the intermediate species (pathway a) and minimize the reaction with other maleimide units (pathway b) giving crosslinked polymers (Fig. 10).

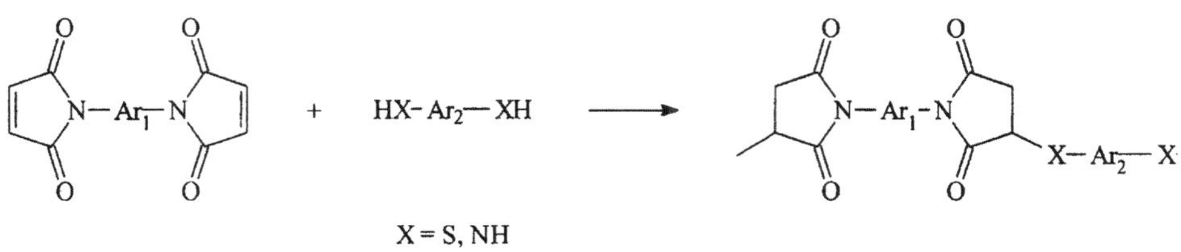

Fig. 9. Step-growth polymerization of BMIs

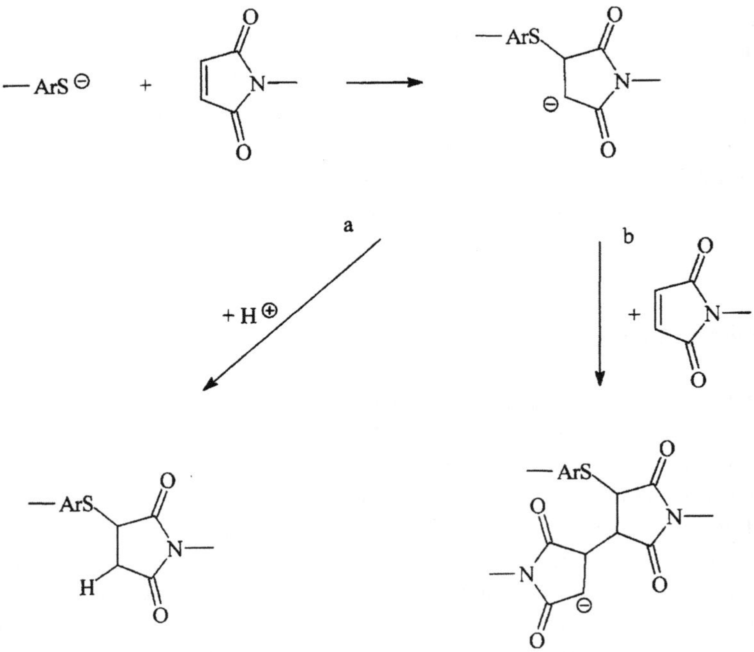

Fig. 10. BMI step-growth polymerization: an acidic medium makes the polyaddition selective

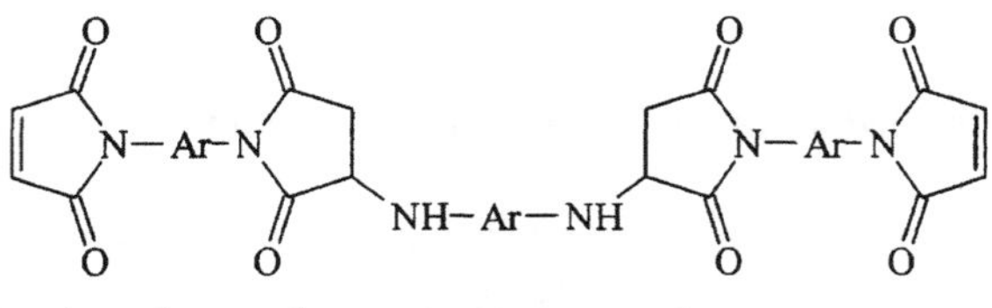

Fig. 11. Bismaleimide end-capped aspartimide (Kerimid)

The reaction with aromatic diamines is catalyzed by carboxylic acid and leads to high molecular weight linear polymers [43]. By controlling the molar ratio of the reactants this reaction has been used to produce low molecular weight polyaspartimides end-capped with maleimide (Fig. 11). Such types of oligomers called Kerimid were formerly commercialized by Rhône-Poulenc.

A similar approach was used by the Shell Chemical Company by reaction of the aminobenzhydrazide with a bismaleimide [44] giving an extended BMI (Fig. 12) commercialized under the trade name Compimide.

These modified BMIs were developed in order to improve the fracture toughness of the BMIs by decreasing the crosslinking density. In fact, commercial BMIs are formulated with comonomers, chain extenders, reactive diluents and sometimes with viscosity modifiers [45].

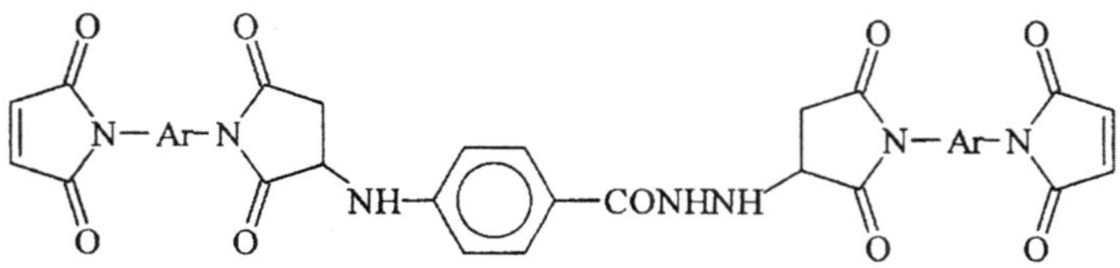

Fig. 12. Maleimide end-capped aminobenzhydrazide (Compimide)

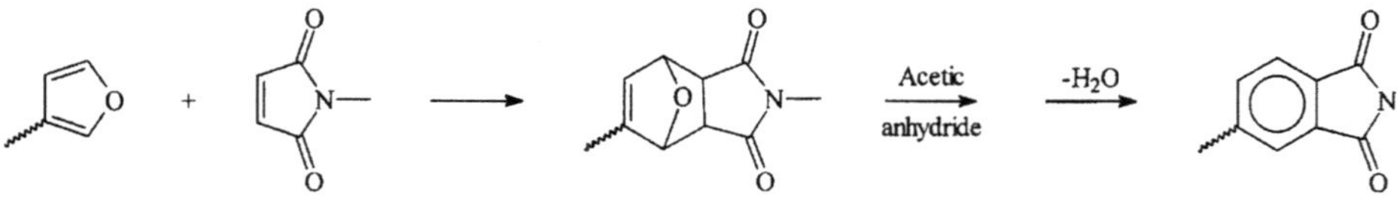

Fig. 13. Diels–Alder reaction of a furan terminated oligomer with a BMI

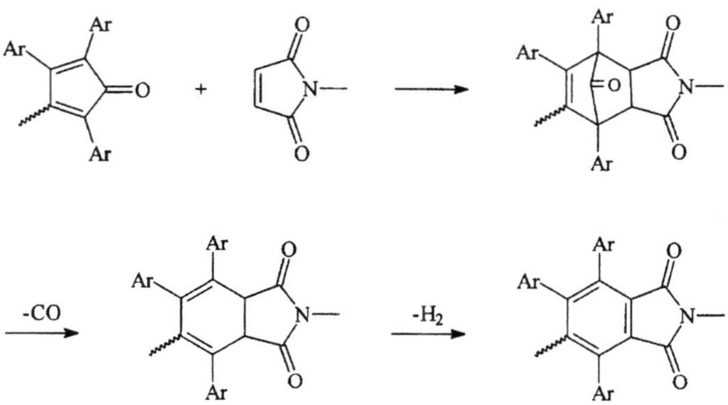

Fig. 14. Synthesis of linear aromatic polyimide by Diels–Alder reaction between a BMI and a biscyclopentadienone

4.2
BMI Diels–Alder reactions

The double bond of maleimides in very reactive towards electron-rich dienes, to give a normal Diels–Alder cycloaddition. Thus BMIs were used to obtain linear polyimides by reaction with several kind of dienes [46–51]. However the dienes are often difficult to prepare [51] and functionalized dienes have been used. Furan terminated oligomers react with BMIs at 70 °C leading to an oxygen-containing cycloadduct [52–57] which can react with acetic anhydride to give an aromatic imide (Fig. 13) [58–59].

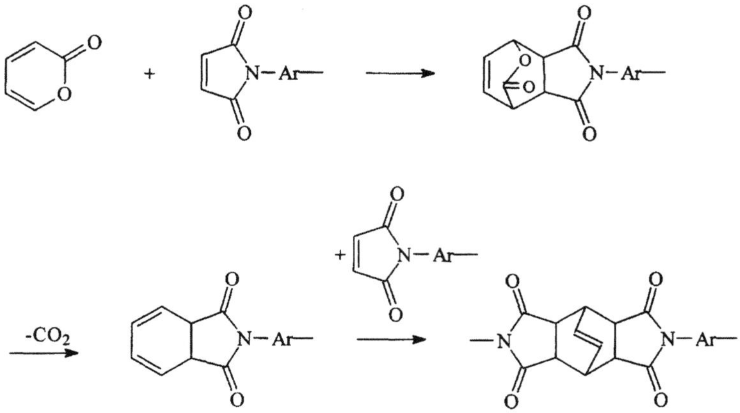

Fig. 15. Bicyclo[2.2.2]octene units containing polyimide obtained by Diels–Alder reaction of a BMI with 2-pyrone

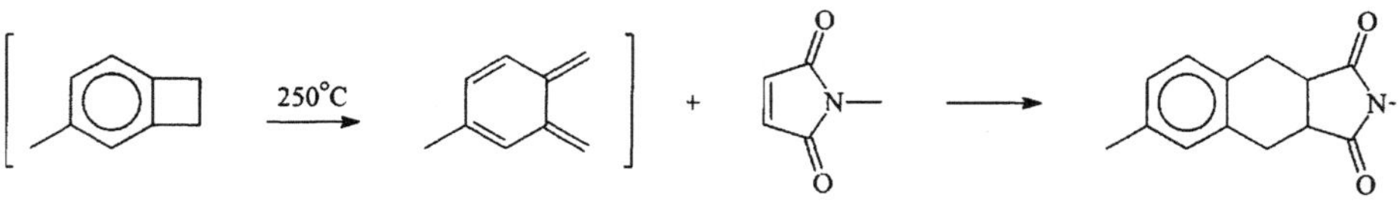

Fig. 16. Chemistry of benzocyclobutene-BMI polyaddition

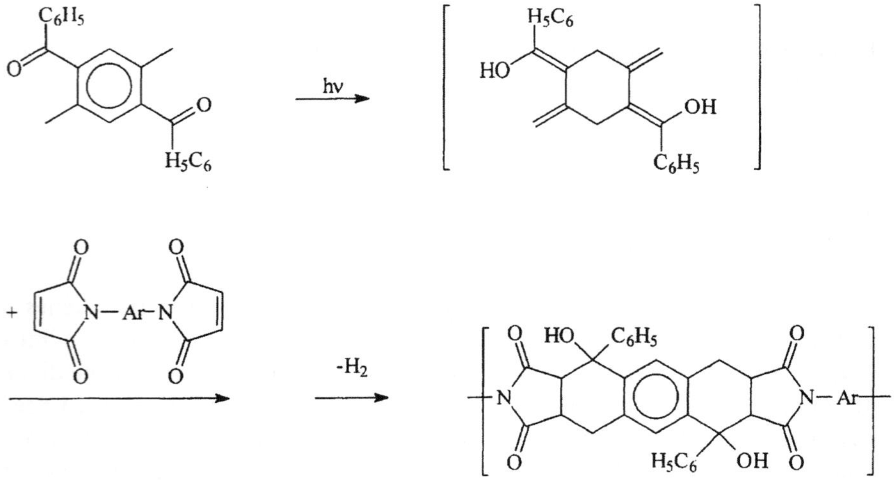

Fig. 17. Diels–Alder polyaddition of a BMI with a transient photoenol

The reaction of a bis-arylcyclopentadienone with a BMI gives a ketonic adduct [60–61]. The carbon monoxide evolution proceeds spontaneously but the final aromatization is then difficult to perform (Fig. 14).

Pyrones behave as dienes and react with bismaleimides giving biscycloadducts [62–65]. By heating, carbon dioxide extrusion takes place with formation of bisdienes. This reaction was used to prepare a polyimide with a bicyclooctene structure [51] (Fig. 15).

When heated, benzocyclobutene reacts as a bis-methylene cyclohexadiene and polyaddition with maleimide has been shown [66] (Fig. 16).

Meador et al. recently reported that a 2,5-diaroyl-*p*-xylene gives a transient photoenol able to react with a BMI giving a poly-1,2,3,4,5,6,7,8-octahydroanthracene cycloadduct [46] (Fig. 17).

4.3 BMI Ene Reactions

BMIs have been copolymerized with unsaturated monomers such as *ortho*-allylphenol [67,68] and *ortho*-propenylphenoxy oligomers. In any case, the first step has been identified as an ene reaction between the allylic or the propenyl double bond followed by a Diels–Alder reaction (Fig. 18). Identification of such a complex structure is very difficult and can be performed only for low conversion ratios. However, the chemistry was intensively studied because the final network can exhibit better mechanical properties than the pure crosslinked nadimides.

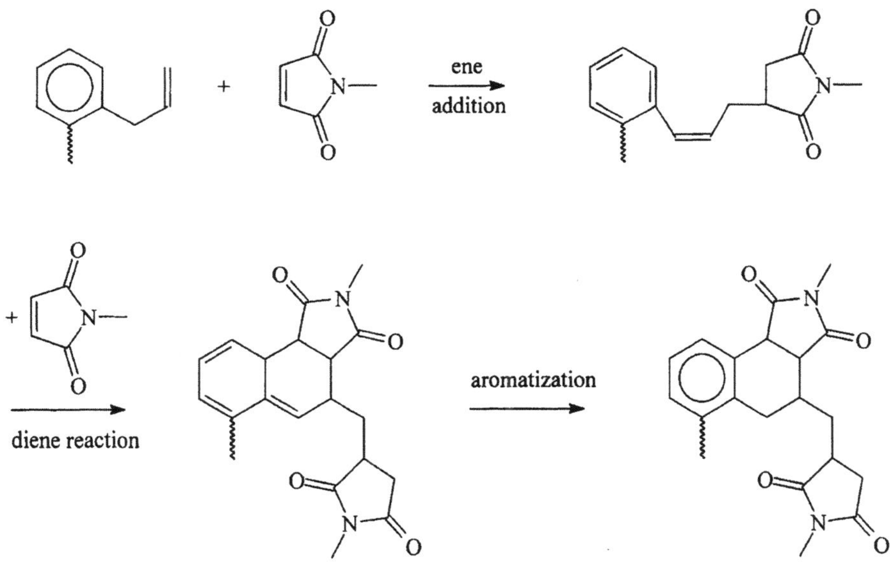

Fig. 18. Ene and diene reactions which take place during the copolymerization of BMI and arylpropenyl oligomers

4.4 Miscellaneous Reactions of BMIs

Hydrosilylation of BMI with 1,4-bis(dimethylsilyl)benzene leads to the formation of polysiloxane polyimide [69].

The reaction of aromatic bis-cyanate with BMI has been investigated. The cyanate trimerizes, giving a cyanurate ring but can also copolymerize with the maleimide double bond giving a pyrimidine ring (Fig. 19) during the cure cycle [70,71].

A donor-acceptor zwitterionic polymerization has been mentioned: 2-ethyloxazoline nucleophilic monomer reacts with the electrophilic *N*-phenyl maleimide giving a 1:1 adduct which polymerizes upon heating (Fig. 20). A similar reaction was observed between bismaleimide and bis-oxazoline giving a crosslinked network stable up to 300 °C [72].

The reaction of a BMI with an acetylenic end-capped oligomer has been undertaken. The mechanism was thought to proceed according to the two pathways indicated in Fig. 21: either by a kind of ene synthesis with formation of an hypothetic cycloallene leading to a dihydronaphthalene derivative (pathway a) or by a condensation concerted with an hydride shift (pathway b). No experimental proof was given for the structure but the condensation of both oligomers gave an expected linear product with a better G_{1c} than the one of BMI alone (324 J m^{-2} vs. 34 J m^{-2}) [73].

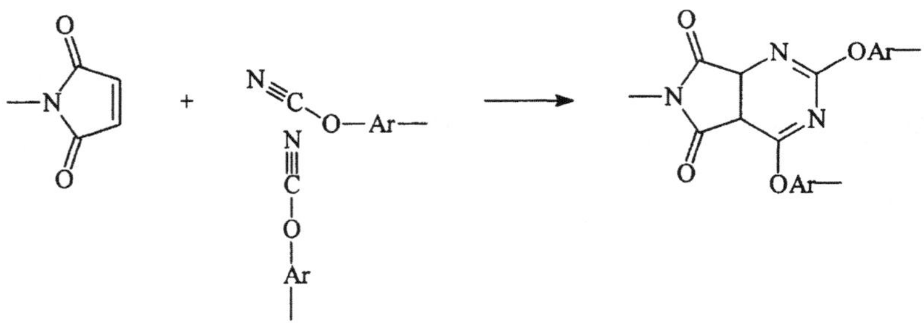

Fig. 19. Suggested mechanism for the copolymerization of BMIs and biscyanates

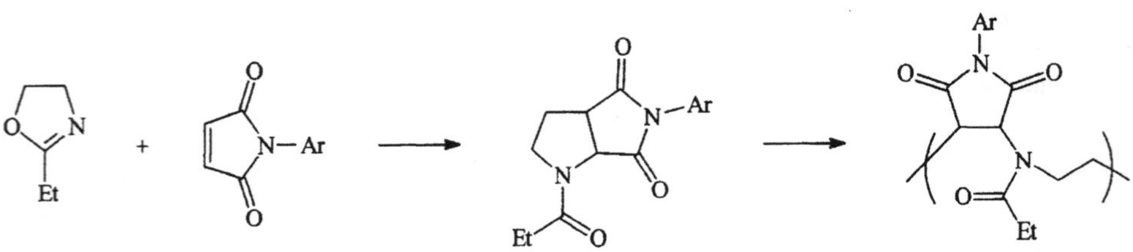

Fig. 20. Zwitterionic polymerization of maleimides and oxazolines

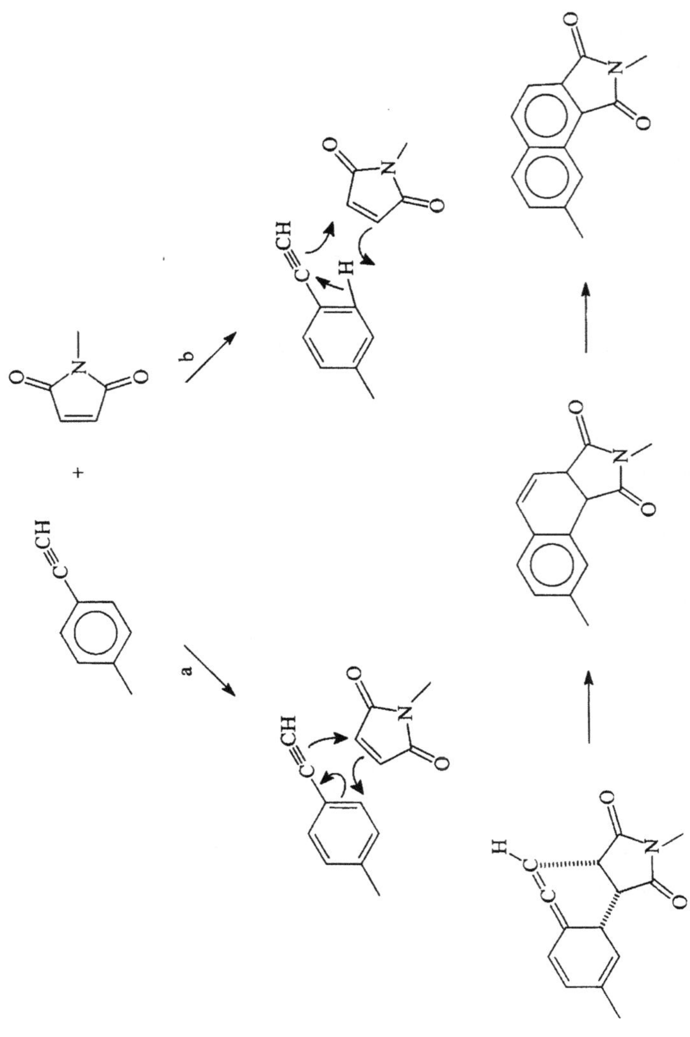

Fig. 21. Suggested pathways for the copolymerization of BMIs and bisacetylenic oligomers

4.5
BMI Thermal Polymerization and Crosslinking

The polymerization of the BMI double bond proceeds by heating the oligomer above the softening point at about 180 °C. The reaction can be initiated with different peroxides as catalyst [74] but also with different kind of amines [75].

Much data is available about the relationships between chemical structure and the processability window (Tp–Tm) determined by DSC. Tp is the temperature onset of exothermic polymerization and Tm is the maximum of the melting temperature curve. However the reliability of these data is questionable because the onset of the polymerization temperature is dramatically dependent on the purity of the BMIs. So, only comparative determinations from the same laboratory should be taken into account.

Another relevant question concerning the BMI polymerization is related to the substitution effect. By DSC determination, taking into account the temperature at which the exothermic heat flow reaches a maximum (Tmax), different authors have compared bismaleimides and biscitraconimides (BCI) (Fig. 22). They found the Tm of BMI higher than the one of the corresponding BCI indicating a higher reactivity for the BCI [76,77]. However, after careful purification with HPLC of a similar BCI and a BMI, Barton et al. [78] pointed out that BMI polymerization takes place at lower temperature than the BCI one. These results are in agreement with the accepted mechanism for a radical polymerization. These authors also made mention of the importance of added impurity which drastically modifies the behavior of BMIs and BCIs, and probably the higher reactivity of BCI mentioned in some papers [76,77] would be due to the presence of the isomeric itaconimide (Fig. 22).

The BMI reactivity is also modified by the presence of electron-withdrawing or -donating groups on the nitrogen substituent. Varma et al. [79] were the first to report this observation by comparing the reactivity of the BMI shown on Fig. 23. Considering the DSC data, taking into account the maximum of the exothermic peak temperature (Tm), it was observed that the Tm value decreases according to the following order : I>II>IV>III. These results show that the reactivity is reduced when a withdrawing group is present.

Working with maleimide end-capped polyether sulfone, Jin and Yee [80] observed a higher E_A for these materials compared with the E_A of BMIs which do not carry a withdrawing group. Their conclusions agree with those of Varma. In addition, these authors mentioned that for the same backbone structure, E_A decreases as the backbone chain length increases, but becomes stable when the prepolymer molecular weight is high enough. This probably means that the chain mobility is also an important factor governing the crosslinking reaction kinetic.

The curing temperature effect on the network nature has been investigated with a model compound of a maleimide end-capped polyether sulfone (Fig. 24).

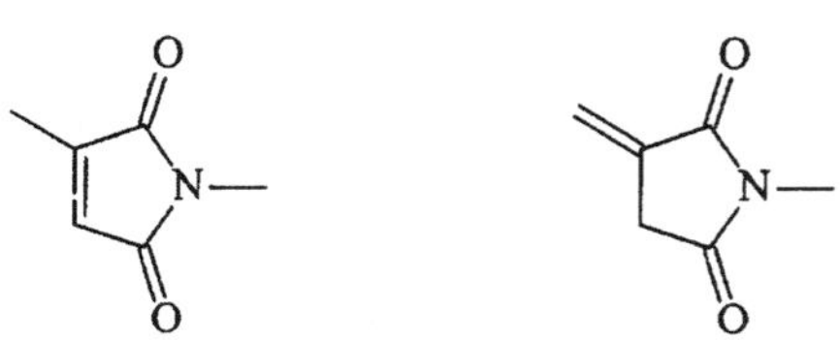

I II

Fig. 22. Structures of *I* Citraconimide and *II* Itaconimide

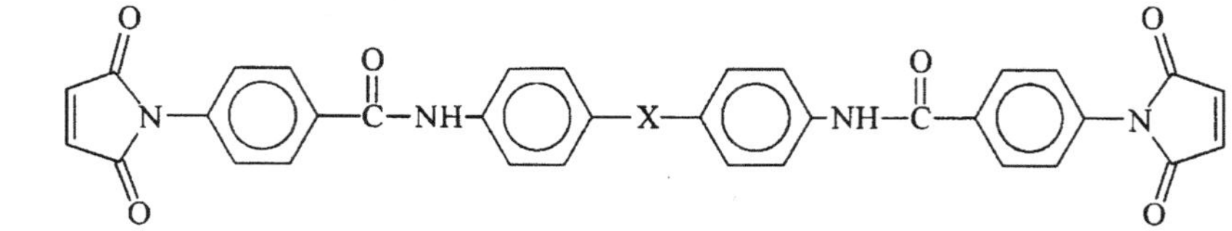

X = —SO_2— I

X = —P(=O)(CH_3)— II

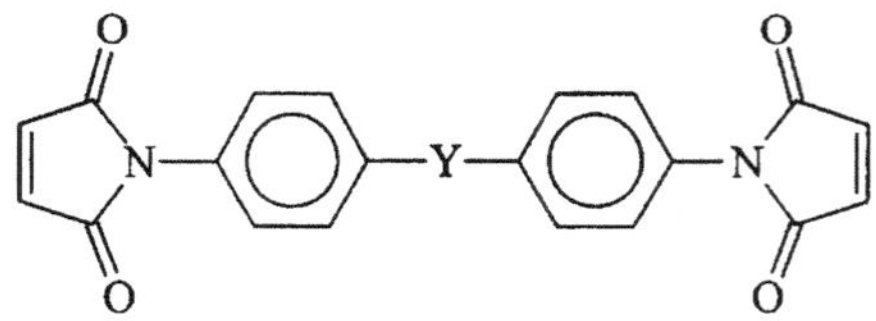

Y = CH_2 III

Y = O IV

Fig. 23. Withdrawing groups containing BMI

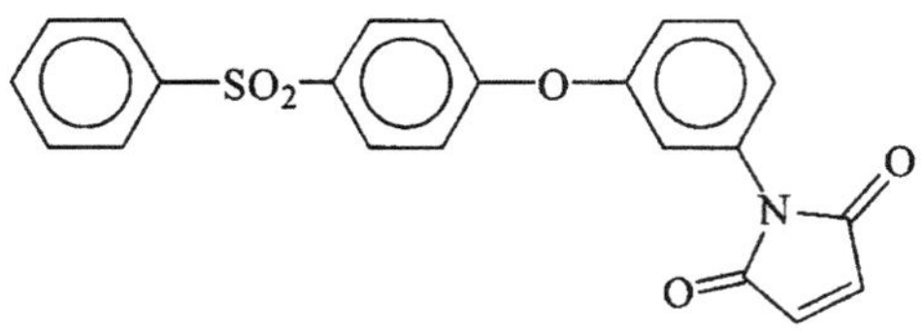

Fig. 24. Model compound which was used to show the trimerization of maleimide

A comparison between curing at 200 and 300 °C has been carried out. The authors showed that the curing at 200 °C gave the highest molecular weight and the highest polydispersity. When the polymerization temperature was increased up to 300 °C the formation of lower molecular weight species (trimer) and a decrease of polydispersity were observed. According to NMR determination the trimer would be a five-membered ring material [81]. A thermal radical depolymerization mechanism could probably take place at high temperature for the polymaleimide. Similar high-temperature depolymerization was mentioned in the case of polyarylacetylene [82].

4.6 BNI Thermal Polymerization and Crosslinking

Upon heating BNIs polymerize to give crosslinked materials formed from the reaction of their ethylenic double bond. As early as the pioneering work of Lubowitz [12] and of a NASA sponsored group [83], a thermal rDA reaction of NAD giving CPD and a MAL unit was evidenced (Fig. 25) and an alternating copolymer of these entities was first proposed (Fig. 26). Shortly after [84], a homonadimide structure was postulated (Fig. 27). The homopolymerization of NAD was thought to be initiated by a radical species, the result of a linear adduct of CPD and MAL.

At the beginning of the 1980s it was suggested [85] that the polynadimide structure is a scrambled copolymer of imides units (Fig. 28a) such as maleimide, nadimide and nadimide-CPD cycloadducts. Thus the CPD is not contributing directly to the polymerization process. These conclusions were supported by the work of a BP group [86]. However, this group also suggested that under prolonged heating the crosslinking process does not stop at this stage. A thermal cleavage of the C_1–C_7 bond (see Fig. 25) of a norbornane bridge is also involved. After loss of a hydrogen radical (Fig. 28b), further crosslinkings give other branchings (Fig. 28c).

Quite recently [41] from solid NMR data of ^{13}C isotope labelled samples of crosslinked PMR-15, the homopolynadimide structure (Fig. 27) was confirmed; but the suggested initiation step is the formation of a diradical generated by the homolytic cleavage of the C_1–C_2 bond of a nadimide unit. At most 15% of the

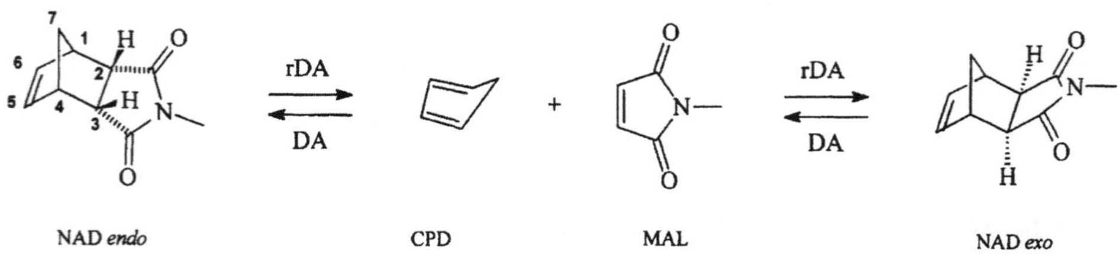

Fig. 25. Nadimide Diels–Alder and reverse Diels–Alder reactions

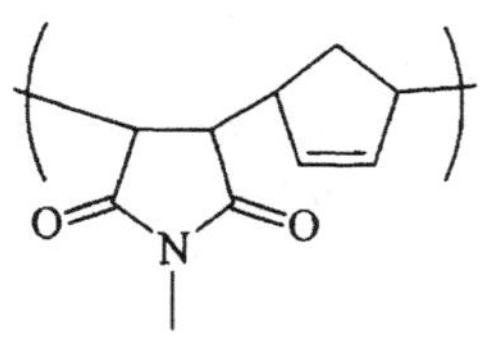

Fig. 26. First structure proposed for the nadimide network [12]

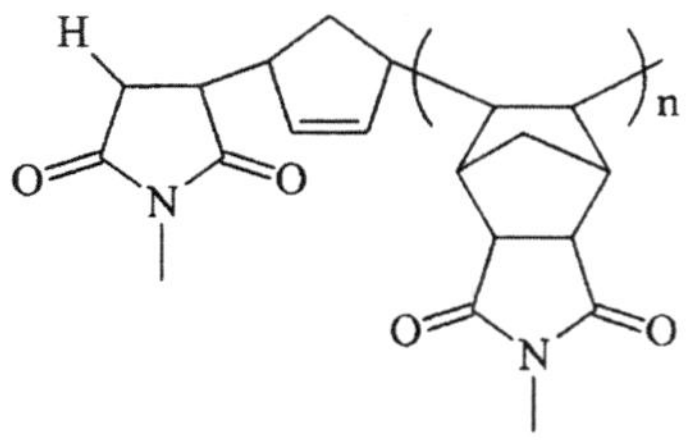

Fig. 27. Nadimide radical homopolymerization initiated by maleimide [84] and cyclopentadiene [41]

PMR-15 crosslinked network could present another structure, which possibly originates from a rDA reaction.

Another approach for determining the structure of the nadimide network and the mechanism of the crosslinking was based on a mass spectrometry study conducted on the thermal treatment product of a model compound [42]. The new outstanding features are:
- each individual macromolecule polymer contain, at most, one maleimide unit,
- polycycloadducts of CPD and NAD were found in all polymerization degrees, and
- some disproportionation products of the polymerized NAD were observed.

To explain these results, an atypic polymerization mechanism is proposed (Fig. 29). Two concomitant reactions, originating from a nadimide rDA one, take place:
- a radical copolymerization initiated by a maleimide unit,
- a mass increased reaction which is the result of a DA cycloaddition of CPD with the unsaturated moiety of imides.

Although this work was carried out with a model compound, the results are quite consistent with the mechanisms suggested by the BP team [86] which was working within a bisnadimide. However, the work of Meador et al. [42] presents some diverging conclusions by minimizing the role of the reverse Diels–Alder reaction in the crosslinking mechanism.

Can the applied pressure modify the nadimide polymerization mechanism? This is a relevant question.

It is well established [87] that pressure accelerates the Diels–Alder reaction and of course the reverse effect is expected for the reverse Diels–Alder reaction. For example, to obtain a homogeneous disk of neat resin or a composite without void the experimental procedure [88] shown on the Fig. 30 has to be used.

The first heating ramp up to 250 °C is carried out without pressure and, therefore, not only solvent removal but also a rDA reaction can occur with initiation of the crosslinking by the maleimide. When the temperature reaches 250 °C, the pressure is applied limiting the rDA during the final step of the polymerization. Moreover, if the nadimide polymerization is conducted without pressure an expanded

a

b

c

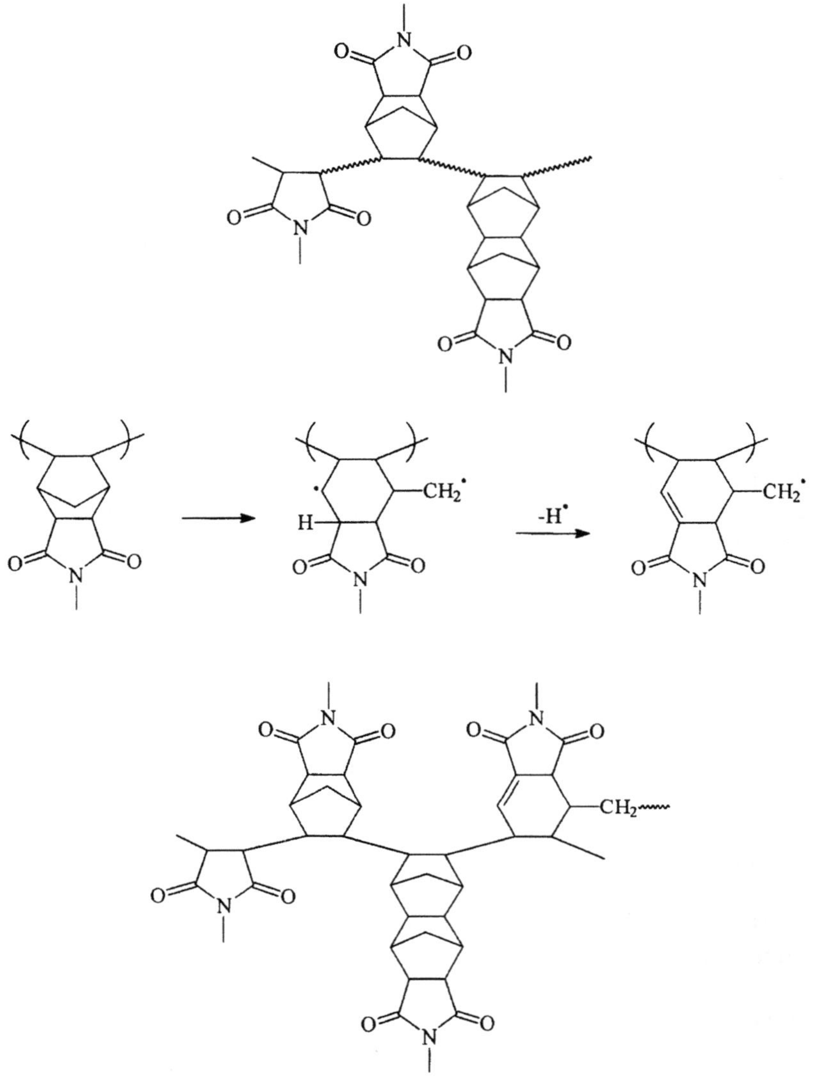

Fig. 28. Mechanism for the nadimide crosslinking [86]

material is obtained. It was demonstrated that the foaming agent is the CPD [32]. These results are consistent with an initiation by the maleimide [89], but does a nadimide double bond give a radical species able to propagate the reaction?

The comparison between the thermal curing of a nadimide and a 1,2,3,6-tetrahydrophthalimide [84,90,91] is interesting. The THPI reacts at higher temperature. A radical initiation takes place which gives only 50% of oligomeric products and an important amount of isomerization and disproportionation products. In the case of the nadimide, a radical mechanism has been evidenced only

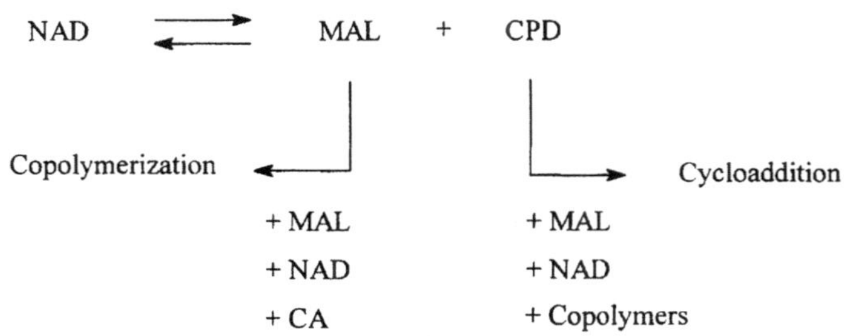

Fig. 29. Main features of nadimide polymerization [42]

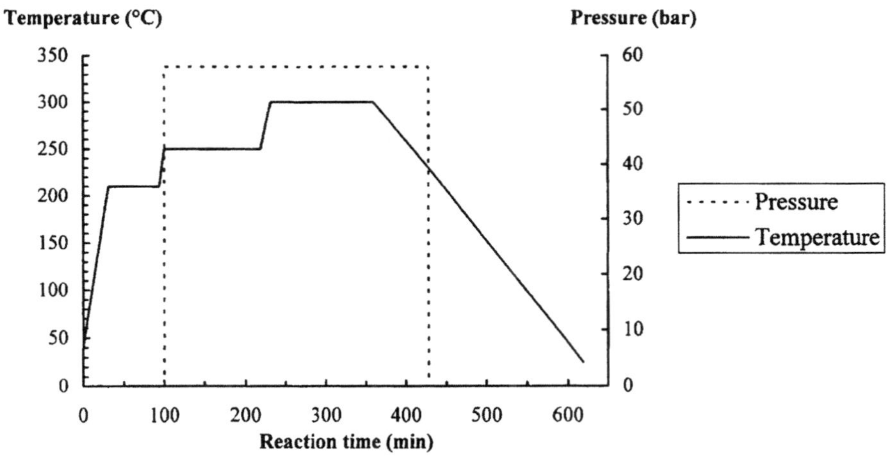

Fig. 30. Nadimide cure cycle for the preparation of molded specimen or composites

when the reverse Diels–Alder reaction was partly or totally inhibited by a substitution on the norbornene ring [92–95].

In conclusion, although the old question concerning the diradical or concerted process of the rDA is still under investigation [96], it seems that the initiation of the NAD polymerization is in any case the rDA reaction.

5 Relationship Between Structure of BMI and BNI End-Capped Oligomers and Properties of the Network

5.1 BMIs

Two types of crosslinked BMIs have to be considered. Firstly the low molecular weight BMIs prepared with monomeric diamines (or a mixture of several monomeric diamines). Secondly the oligomeric BMIs obtained from amino end-capped oligomers. For the BMIs belonging to the first group, Stenzenberger [1] reported interesting data allowing a comparison between the effect of the chemical structure on the melting temperature and also on the temperature of the polymerization exotherm and on the polymerization enthalpy. For similar structures, *meta* instead of *para* catenation decreases the melting temperature as expected but increases the heat of polymerization. This observation probably means that a more flexible chain allows a higher conversion before vitrification.

The polar group effect on the initial melting point and on the T_g of the crosslinked network has been shown by structure modification as exemplified in Fig. 31. It should be mentioned that the highest T_g observed with the nitrile group is obtained with a lower crosslinking density as can be concluded by comparison between the polymerization enthalpies (ΔH Ar_2=186 kJ mol^{-1}/ΔH Ar_3= 109 kJ mol^{-1}) [97]. Some other relationships are also discussed. For example, introduction of CF_3 decreases the melting point of the BMIs and decreases the permittivity of the corresponding network [98].

Different teams have been working on maleimide terminated oligomers (second group). Most of the works have been devoted to the polyether sulfone systems mainly for the following reason: although polyester sulfones are interesting engineering thermoplastics with high T_gs and good fracture toughness, their solvent resistance need to be improved. It was expected that a crosslinking by terminal maleimide groups should increase the T_g and should decrease the swelling which takes place by the solvent attack.

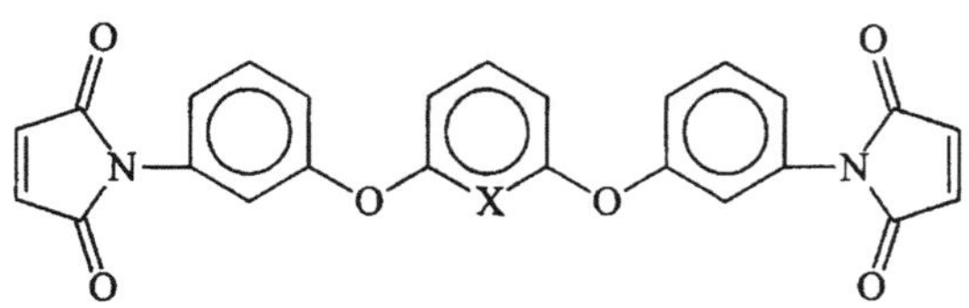

		Mp (°C)	Tg (°C)
Ar_1	X= CH	116	372
Ar_2	X= N	137	386
Ar_3	X= C(CN)	205	404

Fig. 31. Polar group effect on the physical properties of BMIs

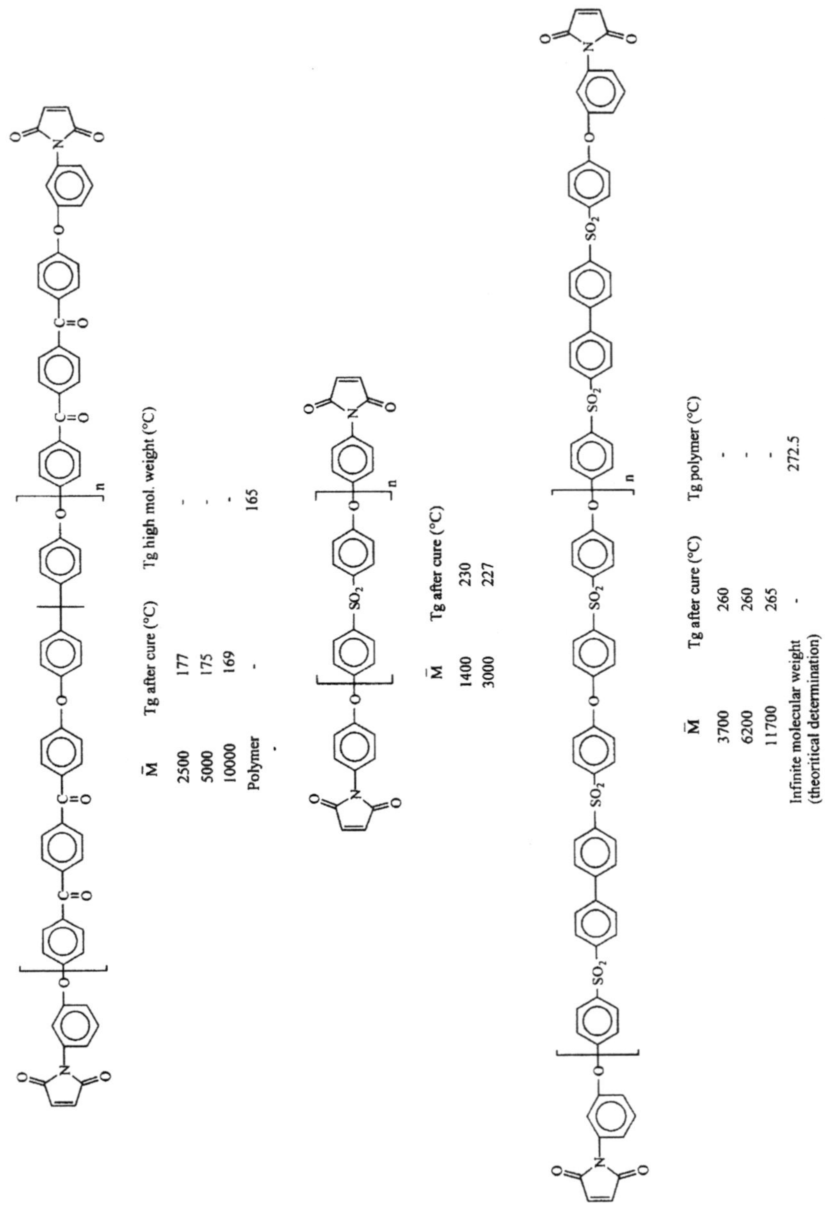

Fig. 32. Variations of glass transition temperature as a function of the molecular weight

As a matter of fact, considering several series of maleimide end-capped oligomers with molecular weight ranging from 1000 to 10,000 Da, it has been observed that the T_g of the network decreases as the molecular weight of the starting oligomer increases. Figure 32 summarizes the results obtained with a polyether ketone [16], a polyether sulfone and a biphenyl containing a polyether sulfone [99].

The conclusions of these works are very similar: firstly the crosslinking mechanism is stopped when the vitrification takes place, secondly the T_g values of the networks are mainly dependent on the nature of the starting end-capped oligomer.

5.2
BNIs

A Mach 2,4 flying airplane would be exposed to –54 °C and 177 °C twice per flight and the expected total life would be 72,000 h [100]. Although PMR-15 is currently used for the manufacture of high-performance composite, it has been evidenced that these laminates show microcracking upon thermal cycling and loss of mechanical properties after long-term ageing. The main drawbacks of the PMR precursors and crosslinked resins are clearly identified:

- presence of free diamine in the precursor leading to a potential toxicity of the prepreg,
- unsatisfactory thermo-oxidative long-term stability, and
- low fracture toughness of the networks.

Many attempts have been devoted to overcoming these shortcomings: improvements to both the first and the second point were expected by chemical modifications but for the third point the most efficient approach was based on blending with linear polymers.

The toxicity of PMR-15 is due to the free 4,4'-methylene dianiline and its replacement by a non-toxic diamine has been one approach. 3,4'-Oxydianiline was introduced into the PMR formulation [101]. 2,2-Bis(4-[4-aminophenoxy]phenyl)hexafluoroisopropane (4-BDAF) was used by BP [102]. The use of this diamine containing four benzenic rings linked by two oxygen atoms and an hexafluorohexapropylidene group decreases the crosslinking density so the T_g is lower than that of crosslinked PMR-15 and the thermal oxidative stability is increased.

The second approach to decreasing the toxicity was to use fully cyclized bisnadimide oligomers. It has been shown that the reaction of a benzhydrol-tetracarboxylic derivative instead of BTDA (Fig. 33) drastically increases the solubility of the fully cyclized oligomer [103]. This oligomer, so called BBN 1500, is soluble at high concentration (50% by weight) in NMP or diglyme and does not exhibit any mutagenic behavior [32]. After crosslinking the thermomechanical characterization performed with a RDA 700 rheometer shows transitions at 325, 68 and –102 °C [35].

Table 4. Thermo-oxidative stabilities of modified PMRs

Resin	T_g (°C)	Loss of weight (%) after 300 h at 188 °C
PMR-15 (control)	370	18
PMR-30	365	12
PMR-50	363	13
PMR-75	358	10

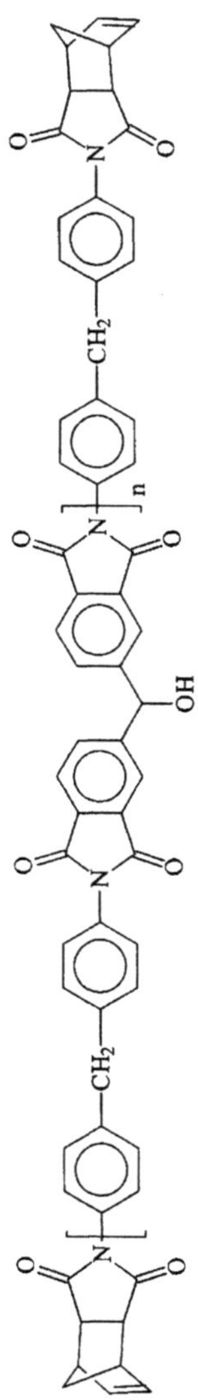

Fig. 33. Fully cyclized soluble benzhydrol bisnadimide (n=2.075, $\bar{M}n$=1500 Da)

Chemical modifications were also undertaken to increase the thermo-oxidative properties of the crosslinked network.

Molecular weight increase is performed by modification of the molar balance between the reactants so BTDE, MDA, NE and PMR-30, PMR-50, PMR-75 with respective theoretical $\bar{M}_n$s of 3000, 5000, 7500 Da were prepared and compared (Table 4) [104]. The T_g of the crosslinked material decreases as the $\bar{M}_n$ increases but the oxidative stability is better for the high molecular weight resins.

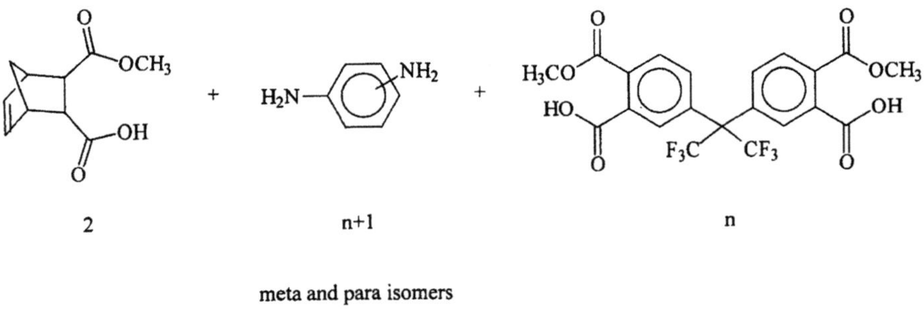

Fig. 34. PMR II chemistry

Table 5. Thermo-oxidative stabilities of modified PMR II

Resin	Theoretical $\bar{M}_n$ (Da)	T_g[1] (°C)	Loss of weight (%) in air[2]
PMR II 13	1300	381	13
PMR II 30	3000	368	8
PMR II 50	5000	355	5

[1]After 24 h post cure at 371 °C. [2]After 300 h under 1 atm.

It has been established by CPMAS ^{13}C NMR and FTIR that the degradation in air of crosslinked nadimide systems proceeds by the oxidation of the cycloaliphatic groups. However the methylene group is also rapidly transformed in an oxidative atmosphere into carbonyl groups [105].

By using hexafluoroisopropylidene diphthalic tetraacid dimethyl ester (HFDE) instead of BTDE and a mixture of *meta*- and *para*-phenylene (mPDA, pPDA) instead of MDA a new generation of more stable nadic resins has been developed (Fig. 34) [106]. Table 5 summarizes the T_gs and the thermo-oxidative properties of these materials as a function of their molecular weight.

It appears that the nadimide end-cap is the main contribution to the thermal degradation of crosslinked PMR. This is confirmed by a comparison between two similar fluorinated PMRs with two different end-capping groups: nadimide or styrene. The styrene end-capped system exhibits the best thermo-oxidative stability.

The hydrolytic stability of crosslinked nadimides has been investigated. It was concluded that moisture absorption is responsible for a reversible plastification effect. However after long-term cycling, the T_g decrease becomes non-reversible due to a chemical degradation [107].

5.3 Blends of BNIs and BMIs with Linear Polymers: Semi-interpenetrated Network Concept

5.3.1 *Introduction*

The values of the stress intensity factor (K_{1c}) and of the strain energy release rate (G_{1c}) of both crosslinked maleic and nadic oligomers are rather low and explain the poor mechanical properties of these materials.

The first approach to overcoming these shortcomings was the chemical modification of both MDA BMI and PMR-15. As a matter of fact introducing more flexible segments between the end-capping groups can increase the toughness. But a detrimental effect on the thermomechanical properties, due to decrease of the glass transition (Tables 4 and 5), was observed.

The concept of interpenetrating the polymer network was introduced in the early 1960s [108]. The basic idea is the formation of blends with two different independent polymer networks on the nano scale. The non-miscibility between two polymers is the general rule and an important question is to know if the gelation takes place before or after the phase separation, because the timing for these two phenomena will govern the size of each network domain [109,110].

When the blend is prepared with a linear polymer and a crosslinkable monomer, a semi-interpenetrated polymer (semi-IPN) network is obtained and this concept has been applied to a linear heterocyclic polymer and a crosslinkable thermostable oligomer [111]. The linear heterocyclic polymers exhibit high glass transition temperatures, good fracture toughness (Table 6), but the high viscosity above T_g make them difficult to process.

Table 6. Thermomechanical properties of some linear polymers used for the preparation of semi-IPNs with BMIs and BNIs

Polymer type	Name	T_g (°C)	G_{1c} (J m^{-2})
PES	Victrex	225	3500
polyether imide	Ultem	217	2500
polyimide ketone	LARC TPI	264	1700
polyimide sulfone	$PISO_2$	273	1400
fluorinated polyimide	Avimide N, NR 150 B2	350	2550
polyamide-imide	Torlon	275	3400

Table 7. End-capped oligomers used for the preparation of thermostable semi-IPNs

Resin type	Name	T_g (°C)	G_{1c} (J m^{-2})
unmodified BMI-MDA	-	300	25
modified BMI	Compimide 793/TM123	280	176
bisnadimide	PMR-15	340	87

The BMIs and BNIs which were used for the semi-IPN preparations are summarized in Table 7. By heating, the melt viscosity of such oligomers decreases before the crosslinking and allows the process with classical tools. Two questions arise: Is it possible to get an intimate mixture of crosslinked BMI or BNI with a linear heterocyclic polymer and what will be the behavior of such semi-IPNs?

5.3.2
Bismaleimide-Linear High T_g Thermoplastic Semi-IPNs

The first question to be discussed is the polymer miscibility which governs the blend morphology. The solubility parameters of BMIs is 12–135 (cal cm^{-3})$^{0.5}$ vs. 11–12 (cal cm^{-3})$^{0.5}$ for the high-performance thermoplastics [112]. We can expect an important non-miscibility; however, the morphology will also depend on some other factors (conversion at the gel point, viscosity...) and as a result different types of morphologies were identified.

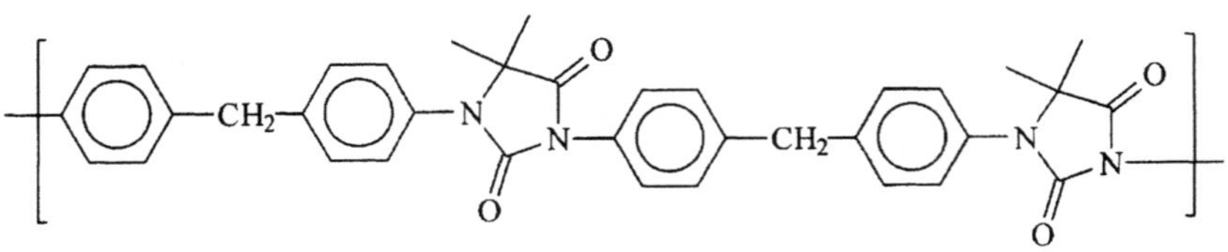

PH 10

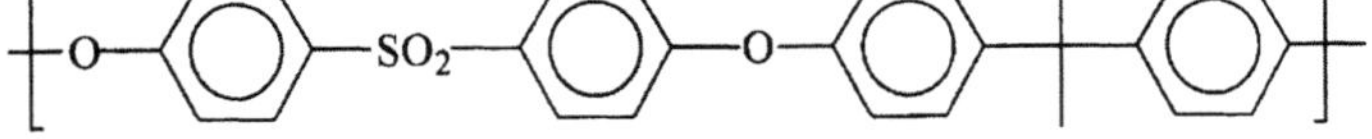

Udel

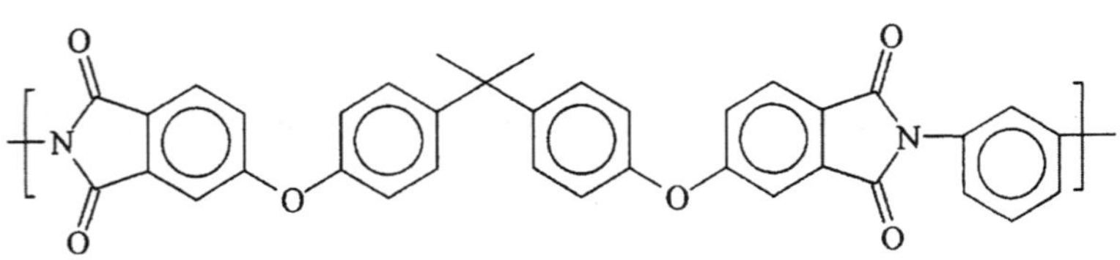

Ultem

Fig. 35. Linear heterocyclic polymers blended with BMIs

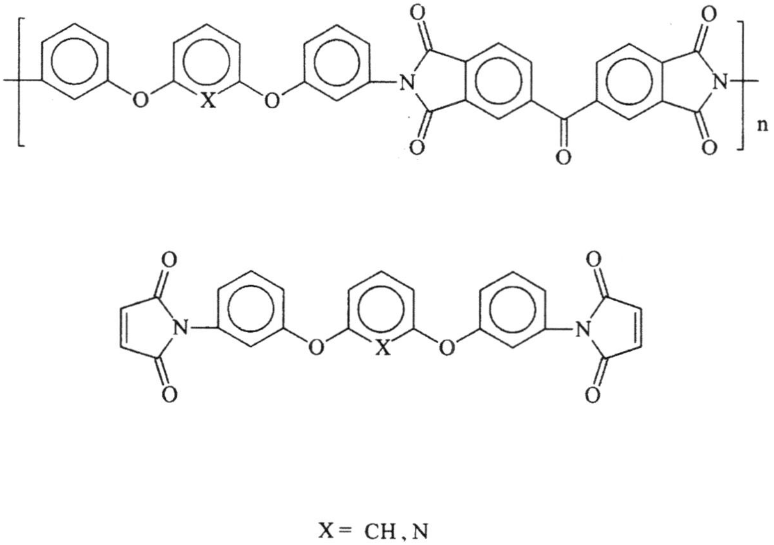

Fig. 36. BMIs and linear polyimides giving partial miscibility after curing [116]

A mixture of a BMI (Compimide 796) with 4,4'-bis(*ortho*-propenylphenoxy)benzophenone (TM 23) was blended in solution with Udel 700 (polysulfone from Union Carbide), Ultem 100 (General Electric polyether imide), and PH 10 (Bayer polyhydantoin) (Fig. 35). The thermoplastics were introduced at various concentrations up to 33%. A phase segregation did not appear with PH 10, but two phases were observed with Ultem. With both semi-IPNs the observed G_{1c} were found to be four- to fivefold the G_{1c} of the neat BMI [113].

The molecular weight of the thermoplastic affects the flexural properties and the G_{1c}. For example, a molecular weight ranging from 30,000 to 50,000 Da for an aromatic polyether gave the best mechanical properties improvement for a blend with Compimide 796–TM 123 [114].

The blending of a BMI with a powder of a thermoplastic polymer without solvent offers a great advantage for the processing. Some results have been published based on a non-disclosed formula BMI blended with the CIBA GEIGY Matrimide 5218. The addition of 20% of powder did not require a modification of the cure cycle and an improvement in the mechanical properties (inhibition of crack propagation) was observed [115].

A partial miscibility between a linear polyimide and a crosslinked BMI was evidenced when the polyimides and the BMIs were prepared with the same diamine (Fig. 36) [116]. As a consequence the adhesive properties of the blend were better at high temperature than the ones of the linear polymer alone.

A similar concept based on a mixture of BMI–MDA (Matrimide 5292 A) and bis-alkenylphenol (Matrimide 5292 B) with a flexible polyimide has been patented as a heat-resistant adhesive [117].

An increase in the K_{1c} and G_{1c} of BMI–MDA was again observed by blending with Ultem 1000. After crosslinking the semi-IPN exhibited phase separation and a complex morphology [118].

5.3.3
Bisnadimide-Linear High T_g Thermoplastic Semi-IPNs

As shown in a previous section, fluorinated nadimides exhibit the best thermo-oxidative behavior. NR 150 linear polymers are prepared by reaction of HFDE and a mixture of *para*- and *meta*-phenylenediamine. A semi-IPN was prepared by addition of a 150 °C staged PMR-15 to a solution of the NR 150 precursor. The solution was used for the composite manufacture. After curing at 250 °C, the DSC diagram showed two peaks in agreement with a non-miscible system. After curing, the G_{1c} of the blend is higher than that of pure PMR-15 (Table 8) [119].

LARC TPI is a linear polyimide prepared with BTDA and 3,3'-diaminobenzophenone. The polyamide acid intermediate was used to prepare a semi-IPN with the PMR-15 reactants. After curing the blend exhibited a twofold T_g and an improved value of G_{1c} (Table 8).

Although the G_{1c}s of the networks are higher than the G_{1c}s of the crosslinked PMR, the observed values are lower than the ones expected according to the rule of mixture [120].

The role of the miscibility of semi-IPN components on the mechanical properties has been discussed. The linear bisnadimide was a benzhydrol bisnadimide (Fig. 33). Three polyimides prepared from the same diamine and three different dianhydrides (Fig. 37) were used as linear components. The blends were cured up to 300 °C in a similar fashion to the bisnadimide alone. The results for the blend containing 20% by weight of linear polymers are summarized in Table 9. The non-miscible character of the components gives a phase segregation leading to the best toughness [121].

The semi-IPN concept has also been used in order to improve the high-temperature adhesive properties of a flexible polyimide. Polyimide prepared with isophthaloyldiphthalic anhydride and *meta*-phenylenediamine presents good adhesive properties but is limited to 270 °C. By blending it with a bisnadimide

Table 8. Thermomechanical properties of semi-IPNs from PMR-15 (T_g=339 °C, G_{1c}= 87 J m^{-2}) after curing

Linear polymer	Semi-IPN
NR 150 B 2	
T_g=352 °C	T_gs=348–321 °C
G_{1c}=2555 J m^{-2}	G_{1c}=368 J m^{-2}
LARC TPI	
T_g=257 °C	T_gs=261–325 °C
G_{1c}=1768 J m^{-2}	G_{1c}=476 J m^{-2}

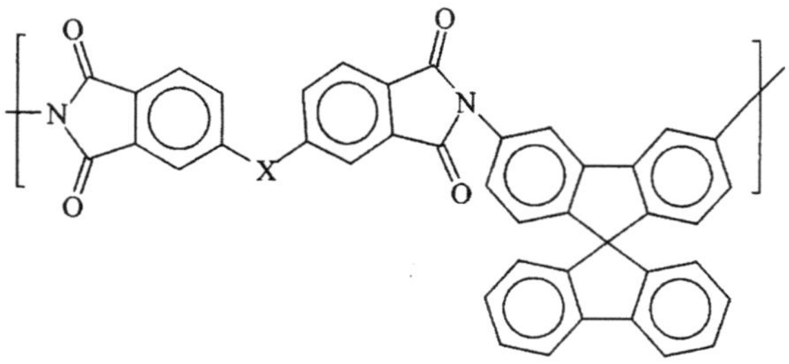

X	Tg (°C)
CHOH	340
CO	350
$C(CF_3)_2$	360

Fig. 37. High T_g polyimides blended with a benzhydrol bisnadimide (Fig. 33) [121]

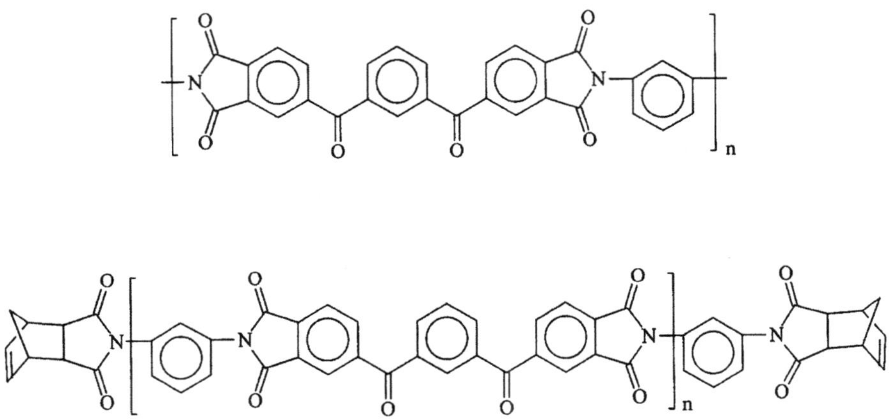

Fig. 38. Linear polyimide and bisnadimide obtained with isophthaloyl diphthalic anhydride and *meta*-phenylenediamine

Table 9. Some properties of BBN 1500 based semi-IPNs

	T_{g1} (°C)	T_{g2} (°C)	K_{1c} (Mpa $m^{0.5}$)	Aspect
neat bisnadimide	271	–	0.89	transparent
20% Polymer containing CHOH[1]	283	–	0.91	transparent
20% Polymer containing CO[2]	285	315	1.06	overcast
20% Polymer containing CF_3[3]	279	320	1.23	opaque

[1]Linear polymer from BHTDA. [2]Linear polymer from BTDA. [3]Linear polymer from HFDA.

obtained from the same starting materials (Fig. 38) a miscible blend was obtained after crossslinking for which the adhesive properties are maintained up to 320 °C [122].

6 Degradation of the BMI and BNI Networks

6.1 BMIs

The thermal degradation of aliphatic and aromatic crosslinked bismaleimides has been studied by thermogravimetry and pyrofield ion mass spectrometry by Stenzenberger et al. [123]. They pointed out that aliphatic BMIs show the lowest stability due to the cleavage of aliphatic bridges giving many fragments containing succinimide or maleimide groups.

The decomposition of aromatic BMIs follows another pathway involving succinimide ring cleavage followed by carbon monoxide evolution. Moreover it must be pointed out that at temperature as low as 260 °C a depolymerization process can occur probably depending on the chain mobility as was evidenced for model compounds [81].

6.2 BNIs

Ageing and thermal degradation of crosslinked nadimide end-capped oligomers have been studied by means of solid infrared ^{13}C NMR spectroscopy as well as by thermogravimetric measurements. They are mainly related to PMR and BBN families.

According to recent results [105,124–126] it seems that three separate degradation stages have been pointed out. The first one, which is the lower temperature process, is the main source of weight loss; it could correspond to the rapid alteration of norbornenyl moeities [42,105,124–127]. Formation of CPD was evidenced [126] and it is more probably formed from polycycloadduct imides than from unreacted nadimides [42].

Moreover the opening of the norbornene bridge was also proposed to form aromatic and hydrogenated moieties through disproportionation reactions [21,89]. The second stage arise from the oxidation of the methylene group. It was accurately demonstrated [124] from ^{13}C NMR of selectively ^{13}C-enriched materials that the oxidation did not stop the formation of a carbonyl group [36,126,128] but it continued to produce also ester and acid entities [124]. The third stage, which is the slowest one, is the degradation of the imide backbone [124,126]. The products of higher-temperature degradation have been identified [129,130], but not correlated with reaction mechanisms.

The crack formations in neat resin specimens was studied during the thermal ageing. The material changes were monitored by optical and electron microscopy and metallography [131] and recently by acoustic emission [132]. Ageing time and temperature are the main factors originating the cracks due to interactions between voids and stresses that develop mainly on the surface.

On the other hand, water uptake is a continuous problem for polyimides and particularly for polynadimides [133]. Dynamic mechanical spectrometry (viscoelastic measurements) have been used to investigate the network degradations due to hydrolytic process [134]. It was shown on a special PMR resin (AFR 700 B) that network reformations are possible through post-curing. In addition, for fluorinated systems, ^{19}F NMR can be used to follow the hydrolysis of the imide groups [135].

7 References

1. Stenzenberger HD (1994) Addition polyimides. In: Hergenrother PM (ed) Advances in polymer science, vol 117. High performance polymers. Springer, Berlin Heidelberg New York, p 163
2. Lin SC, Pearce EM (1994) High-performance thermosets: chemistry, properties, applications. Hanser, Munich
3. Sillion B (1973) Polyimides and other heteroatomic polymers. In: Allen G, Bevington JC, Eastmond GC, Ledwith A, Russo S, Sigwalt P (eds) Comprehensive polymer science, vol 5, The synthesis, characterization, reactions and applications of polymers. Pergamon Press, Oxford, p 499
4. Sillion B, Rabilloud G (1995) Heterocyclic polymers with high glass transition temperatures. In: Ebdon JR, Eastmond GC (eds) New methods of polymer synthesis. Blackie Academic and Professional, London, p 246
5. Chandra R, Rajabi L (1997) J Macromol Sci, Rev Macromol Chem Phys C37:61
6. Jin S, Yee AF (1991) J Appl Polym Sci 43:1849
7. Lyle G, Hedrick JC, Lewis DA, Senger JS, Chen DH, Wu SD, McGrath JE (1989) Synthesis and characterization of maleimide terminated poly(arylene ether sulfone)s. In: Feger C, Khojasteh MM, McGrath JE (eds) Polyimides: materials, chemistry and characterization. Elsevier, Amsterdam, p 213
8. Sauer CK (1969) J Org Chem 34:2275
9. Endrey AL (1965) US Patent 3,171,631
10. Winter H, Loontjens JA, Mostert Ham, Tholen MGW (1989) Chemistry and properties of new bismaleimides designed for improved processability. In: Feger C, Khojasteh MM, McGrath JE (eds) Polyimides: materials, chemistry and characterization. Elsevier, Amsterdam, p 229
11. Tatsuhiro Y, Masahiro T, Yoshuki G (1997) Jpn Kokai Tokkyo Koho JP 09 12,984 (97 12,984). from (1997) Chem Abst 126:200523e
12. Lubowitz HR (1971) Preprints (Am Chem Soc Div Org Coat Plast Chem) 31(1):561
13. Volksen W (1994) Condensation polyimides: synthesis, solution behavior, and imidization characteristics. In: Hergenrother PM (ed) Advances in polymer science Vol 117 High performance polymers. Springer, Berlin Heidelberg New York, p 111
14. Serafini TT, Delvigs P, Lightsey GR (1972) J Appl Polym Sci 16:905
15. (a) Wilson D, Polyimides as resin matrices for advanced composites.; (b) Hergenrother PM, Polyimides as adhesives.; (c) Young PR, Escott R, Characterization of polyimides. In: (1990) Wilson D, Stenzenberger HD, Hergenrother PM (eds) Polyimides. Blackie, Glasgow, pp 187, 158, 129
16. Lyle GD, Senger JS, Chen DH, Kilic S, Wu SD, Mohanty DK, McGrath JE (1989) Polymer 30:978

17. Grenier-Loustalot MF, Gouarderes F, Joubert F, Grenier P (1993). Polymer 34:3848
18. Scola DA (1985) Some chemical characteristics of the reverse Diels–Alder polyimide, PMR-15. In: Gupta MR, Weber WD (eds) Polyimides: synthesis, characterization and application (Proceedings of the Second International Conference on Polyimides). Society of Plastics Engineers, Inc, Hopewell Jct, p 247
19. Wong AC, Ritchey WM (1980) Spectroscopy Lett 13:503
20. Young PR, Chang AC (1983) J Heterocyclic Chem 20:177
21. Hay JN, Boyle JD, Parker SF, Wilson D (1989) Polymer 30:1032
22. Bertholio F, Mison P, Pascal T, Sillion B (1993) High Perform Polym 5:47
23. Laguitton B, Mison P, Pascal T, Sillion B (1995) Polym Bull 34:425
24. Hay JN, Boyle JD, James PG, Walton JR, Wilson D (1989) Polymerization mechanism in PMR-15 polyimide. In: Feger C, Khojasteh MM, McGrath JE (eds) Polyimides: materials, chemistry and characterization. Elsevier, Amsterdam, p 305
25. Milhourat-Hammadi A, Chayrigues H, Levoy R, Merienne C, Gaudemer A (1991) J Polym Sci Part A Polym Chem 29:1347
26. Milhourat-Hammadi A, Chayrigues H, Merienne C, Gaudemer A (1994) J Polym Sci Part A Polym Chem. 32:203
27. Grenier-Loustalot MF, Grenier P (1991) High Perform Polym 3:263
28. Grenier-Loustalot MF, Grenier P (1991) High Perform Polym 3:113
29. Pascal T, Audigier D, Mercier R, Sillion B (1991) Polymer 32:1119
30. Parker SF, Hoyle ND, Walton JR (1990) High Perform Polym 2:267
31. Laguitton B (1995) PhD thesis, Claude-Bernard-Lyon-I University (n 269–95)
32. Dutruch L, Senneron M, Bartholin M, Mison P, Sillion B (1997) Preparation of thermostables rigid foams by control of thereverse Diels–Alder reaction during the crosslinking of bisnadimide oligomers. In: Khemani KC (ed) Am Chem Soc Symp Series Vol 669 Polymeric foams: science and technology. Am Chem Soc, Washington DC, p 37
33. Kranbuehl D, Hood D, Wang Y, Boiteux G, Stephan F, Mathieu C, Seytre G, Loos A, McRae D (1997) Polym Adv Technol 8:93
34. Stephan F, Seytre G, Boiteux G, Ulanski J (1994) J Non-Cryst Solids 172–174:1001
35. Habas JP, Peyrelasse J, Grenier-Loustalot M.F (1996) High Perform Polym 8:515
36. Habas JP, Peyrelasse J, Grenier-Loustalot M.F (1996) High Perform Polym 8:579
37. Pater RH (1994) SAMPE J 30:29
38. Bounor-Legaré V, Mison P, Sillion B (1997) Polymer 39:2815
39. Parker SS, Doyle ND, Walton JR (1991) Infrared studies of the cure of PMR-15. In: Abadie MJM, Sillion B (eds) Polyimides and other high-temperature polymers (Proceedings of the 2nd European Technical Symposium on Polyimides and High Temperature Polymers, STEPI 2). Elsevier, Amsterdam, p 301
40. Swanson SA, Fleming WW, Hofer DC (1992) Macromolecules 25:582
41. Meador MAB, Johnston JC, Cavano PJ (1997) Macromolecules 30:515
42. Damerval V, Delolme F, Mison P, Sillion B (1996) Thermal polymerization of nadimides: reaction mechanism study by mass spectrometry (LSIMS) In: Abadie MJM, Sillion B (eds) Polyimides and high performance polymers (Proceedings of the 4th European Technical Symposium on Polyimides and High Performance Polymers, STEPI 4). Montpellier-2 University Press, Montpellier, p 169
43. White JE (1986) Ind Eng Chem Prod Res Dev 25:395
44. Stenzenberger HD (1980) US Patent 4,211,861
45. Stenzenberger HD (1988) Br Polym J 20:383
46. Meador MAB, Meador MA, Williams LL, Scheiman DA (1996) Macromolecules 29:8983
47. Smith Jr JG, Sun F, Ottenbrite RM (1996) Macromolecules 29:1123
48. Ottenbrite RM, Yoshimatsu A, Smith Jr JG (1990) Polym Adv Technol 1:117
49. Rienecke M, Ritter H (1994) Macromol Chem Phys 195:2445
50. Kiselev VD, Sakhabutdinov AG, Shakirov IM, Zverev VV, Konovalov AI (1992) J Org Chem SSSR 28:1806

51. Turpin F, Mison P, Sillion B (1996) Synthesis of linear polyimides using Diels–Alder polyaddition. In: Feger C, Khojasteh MM, Molis SE (eds) Polyimides: trend in materials and applcations (Proceedings of the Fifth International Conference on Polyimides). Society of Plastics Engineers Inc, Hopewell Jct, p 169
52. Laita H, Boufi S, Gandini A (1997) Euro Polym J 33:1203
53. Kuramoto N, Hayashi K, Nagai K (1994) J Polym Sci, Part A Polym Chem 32:2501
54. Hawthorne DG, Hodgkin JH, Jackson MB, Morton TC (1994) High Perform Polym 6:249
55. Daikoumakos CD, Microyannidis JA (1994) Eur Polym J 30:465
56. Patel HS, Mathur AB, Bhardwaj IS (1994) Polym Sci 1:39
57. Patel HS, Patel NR (1994) High Perform Polym 6:13
58. Tesoro GC, Sastri VR (1986) Ind Eng Chem Prod Res Dev 25:444
59. Jiang B, Jiang L, Cai X (1996) Chengdu Keji Daxue Xuebao 4:45. From (1997) Chem Abst 126:186 412 j
60. Kuhrau M, Stadler R (1993) Polym Int 31:71
61. Harris FW, Norris SO (1973) J Polym Sci Polym Chem Ed 11:2143
62. Alhakimi G, Klemm E, Görls H (1995) J Polym Sci Part A Polym Chem 33:1133
63. Alhakimi G, Klemm E (1995) J Polym Sci Part A Polym Chem 33:767
64. Alhakimi G, Görls H, Klemm E (1994) Macromol Chem Phys 195:1569
65. Kottner N, Bublitz R, Klemm E (1996) Macromol Chem Phys 197:2665
66. Tan LS, Soloski EJ, Arnold FE (1987) Polym Mat Sci Eng 56:650
67. King JJ, Chaudhary M,. Zahir S (1984) Natl SAMPE Symp Exhib 29:392
68. Stenzenberger HD, König P, Herzog M, Römer W, Pierce S, Canning MS (1987) Int SAMPE Symp Exhib 32:44
69. Vygodskii YS, Kuznetsov VL, Pryakhina TA, Zavin BG, Strelkova TV, Chupochkina NA (1995) Vysokomol Soedin Ser A Ser B 37:1621. From (1996) Chem Abst 125:197 115 v
70. Iguchi F, Ikeguchi N, Takigawa T (1979) Proc Electr Electron Insul Conf 14:344
71. Motoori S, Kinbara H, Gaku M, Ayano S (1981) Proc Electr Eletron Insul Conf 15:168
72. Huang SJ, Wilson DM, Ho LH, Jones KD, Di Benetto AT (1994) Composite Structures 27:25
73. Pater RH (1988) Int SAMPE Tech Conf 20:174
74. Grundschober F, Sambeth J (1968) US Patent 3,380,964
75. Zahir SA., Renner A (1978) US Patent 4,100,140
76. Melissaris AP, Mikroyannidis JA (1988) J Polym Sci Polym Chem Ed 26:1165
77. Varma IK, Fohlen GM, Parker JA (1982) J Polym Sci Polym Chem Ed 20:283
78. Barton JM, Hamerton I, Rose JB, Warner D (1991) A comparative study of the thermal behaviour of some aryl bismaleimides and biscirtraconimides. In: Abadie MJM., Sillion B (eds) Polyimides and other high temperature polymers (Proceeding of the 2nd European Technical Symposium on Polyimides and High Temperature Polymers, STEPI 2). Elsevier, Amsterdam, p 283
79. Varma IK, Sharma S (1985) Polymer 26:1561
80. Jin S, Yee AF (1991) J Appl Polym Sci 43:1865
81. Maes C, Devaux J, Legras R, Parsons IW (1995) J Polym Sci Part A Polym Chem 33:1943
82. Gandon S, Mison P, Sillion B (1996) Mechanism of step-growth thermal polymerization of arylacetylene. In: Hedrick JL, Labadie JW (eds) Step-growth polymers for high performance materials, new synthetic methods. Am Chem Soc Symp Series 624:306
83. Burns EA, Jones RJ, Vaughan RW, Kendrick WP (1970) Thermally stable laminating resins NASA CR-72633
84. Serafini TT, Delvigs P (1973) Appl Polym Symp 22:89
85. Wong AC, Garroway AN, Ritchey WM (1981) Macromolecules 14:832
86. Hay JN, Boyle JD, Parker SF, Wilson D (1989) Polymer 30:1032
87. Pindur U, Lutz G, Otto C (1993) Chem Rev 93:741

88. Hilaire B, Verdu J (1991) Some chemical and physical aspects of the cure of polynadimides of the IP 960 type. In: Abadie MJM, Sillion B (eds) Polyimides and other high temperature polymers (Proceeding of the 2nd European Technical Symposium on Polyimides and High Temperature Polymers, STEPI 2). Elsevier, Amsterdam, p 309
89. Damerval V (1997) PhD thesis, Claude-Bernard-Lyon-I University (n 303–97)
90. Scola DA, Stevens MP (1981) J Appl Polym Sci 26:231
91. Bounor-Legaré V, Mison P, Sillion B (1997) Polymer 39:2825
92. Iratcabal P, Cardy H (1995) J Org Chem 60:6717
93. Wong AC, Ritchey WM (1981) Macromolecules 14:825
94. Sukenik CN, Ritchey WM, Malhotra V, Varde U (1987) The synthesis, characterization and thermal chemistry of modified norbornenyl PMR endcaps. In: Serafini TT (ed) High temperature polymer matrix composites. Noyes Data Corporation, Park Ridge, p 77
95. Panigot MJ, Waters JF, Varde U, Sutter JK, Sukenik CN (1992) Macromolecules 25:530
96. Horn BA, Herek JL, Zewail AH (1996) J Am Chem Soc 118:8755
97. Pascal T, Mercier R, Sillion B (1989) Polymer 30:739
98. Nagai A, Takahashi A, Suzuki M, Muroh A (1992) J Appl Polym Sci 44:159
99. Maes C, Devaux J, Legras R, McGrail PT (1995) Polymer 36:3159
100. Hergenrother PM, Rogalski ME (1992) Polym Prepr (Am Chem Soc Div Polymer Chem) 33(1):354
101. Scola DA (1991). Synthesis and characterization of polyimides for high temperature applications. In: Abadie MJM, Sillion B (eds) Polyimides and other high temperature polymers (Proceeding of the 2nd European Technical Symposium on Polyimides and High Temperature Polymers, STEPI 2). Elsevier, Amsterdam, p 265
102. Hoyle ND, Stewart NJ, Wilson D, Baschant M, Merz H, Sikorski S, Greenwood J, Small GD (1989) High Perform Polym 1:285
103. Malinge J, Garapon J, Sillion B (1988) Br Polymer J 20:431
104. Vannucci RD (1987) SAMPE Q 19(1):31
105. Meador MAB, Lowell CE, Cavano PJ, Herrera-Fierro P (1996) High Perform Polym 8:363
106. Chuang KC, Vannucci RD, Ansari I (1991) Polym Prepr (Am Chem Soc Div Polym Chem) 32(2):197
107. Thorp KEG, Crasto AS (1995) Proc Am Soc Composite Tech Conf (10th) p 601
108. Frisch HL, Frisch KC, Klempner D (1977) Chemtech 7(3):188
109. Sperling LH (1981) Interpenetration polymer networks and related materials. Plenum Press, New York
110. Klempner D, Frisch KC (eds) (1989) Advances in interpenetrating polymer networks. Technomic, Lancaster
111. Pascal T, Sillion B (1994) Heat resistant semi-IPNs. In: Klempner D, Frisch KC (eds) Advances in interpenetrating polymer networks, Vol IV. Technomic, Lancaster, p 141
112. Gisholm MS, McGrail PT (1990) Pro Interdisciplinary Symp San Diego, Jan 22–25
113. Stenzenberger HD, Römer W, Herzog M, König P (1988) Int SAMPE Symp Exhib 33:1546
114. Stenzenberger HD, Römer W, Hergenrother PM, Jensen B, Breitigam W (1990) Int SAMPE Symp Exhib 35:2175
115. Blair MT, Steiner PA, Willis EN (1988) Int SAMPE Symp Exhib 33:524
116. Pascal T, Mercier R, Sillion B (1990) Polymer 31:78
117. Yoshikawa A, Ito T, Takatsudo T, Takeda T, Enoki H (1997) Jpn Kokai Tokkyo Koho JP 09 25,470 (97 25,470). From (1997) Chem Abst 126:226310k
118. Abate M, Martuscelli E, Musto P, Ragosta G (1997) Angew Makromol Chem 246:23
119. Pater RH, Morgan CD (1988) SAMPE J 24(5):25
120. Pater RH, Partos RD (1989) New high performance semi-interpenetrating polyimide networks and composites having improved toughness and microcracking resistance. In: Feger C, Khojasteh MM, McGrath JE (eds) Polyimides materials chemistry and characterization. Elsevier, Amsterdam, p 37

121. Dutruch L, Pascal T, Durand V, Senneron M, Sillion B (1997) Polym Adv Technol 8:8
122. Durand V, Senneron M, Sillion B (1994) Polym Prepr (Am Chem Soc Div Polym Chem) 35(1):365
123. Stenzenberger HD, Heinen KU, Hummel DO (1976) J Polym Sci Polym Chem Ed 14:2911
124. Meador MAB, Johnston JC, Cavano PJ, Frimer RA (1997) Macromolecules 30:3215
125. Meador MAB, Johnston JC, Frimer AA, Cavano PJ (1997) Polym Prep (Am Chem Soc Div Polym Chem) 38(1):827
126. Dutruch L (1996) PhD thesis, Claude-Bernard-Lyon-I University (n 252–96)
127. Sutter JK, Jobe JM, Crane EA (1995) J Appl Polym Sci 57:1491
128. Milhourat-Hammadi A, Gaudemer F, Merienne C, Gaudemer A (1994) J Polym Sci Part A Polym Chem 32:1593
129. Torrecillas R, Baudry A, Dufay J, Mortaigne B (1996) Polym Degrad Stab 54:267
130. Torrecillas R, Regnier N, Mortaigne B (1996) Polym Degrad Stab 51:307
131. Bowles KJ, Jayne D, Leonhardt TA (1993) SAMPE Q 242:2
132. Favre JP, Raud C (1995) AECM-5 Int Symp Acoust Emis Compos Mat 5th p 33
133. Gibbs HH (1996) Hydrothermal effects in polyimide composites. In: Abadie MJM, Sillion B (eds) Polyimides and high performance polymers (Proceedings of the 4th European Technical Symposium on Polyimides and High Performance Polymers, STEPI 4). Montpellier-2 University Press, Montpellier, p 306
134. Lee A (1996) High Perform Polym 8:475
135. Smith CD, Mercier R, Waton H, Sillion B (1993) Polymer 34:4852

Received: March 1998

Author Index Volumes 101–140

Author Index Volumes 1–100 see Volume 100

Subject Index

MIX
Papier aus verantwortungsvollen Quellen
Paper from responsible sources
FSC® C105338

If you have any concerns about our products,
you can contact us on
ProductSafety@springernature.com

In case Publisher is established outside the EU,
the EU authorized representative is:
Springer Nature Customer Service Center GmbH
Europaplatz 3, 69115 Heidelberg, Germany

Printed by Libri Plureos GmbH
in Hamburg, Germany